METHODS IN MOLECULAR BIOLOGY™

T0190133

Series Editor
John M. Walker
School of Life Sciences
University of Hertfordshire
Hatfield, Hertfordshire, AL10 9AB, UK

For other titles published in this series, go to
www.springer.com/series/7651

Inositol Phosphates and Lipids

Methods and Protocols

Edited by

Christopher J. Barker

The Rolf Luft Research Center for Diabetes and Endocrinology, Karolinska Institutet, Stockholm, Sweden

 Humana Press

Editor
Christopher J. Barker
The Rolf Luft Research Center for Diabetes and Endocrinology
Karolinska Institutet
Stockholm, Sweden
chris.barker@ki.se

ISSN 1064-3745   e-ISSN 1940-6029
ISBN 978-1-61779-689-0   ISBN 978-1-60327-175-2(eBook)
DOI 10.1007/978-1-60327-175-2
Springer New York Dordrecht Heidelberg London

Preface

This book seeks to cover a broad range of techniques encountered by those working within the field of phosphorylated inositols. This field started with the observation by the Hokins of the turnover of phosphatidylinositol, a minor membrane lipid, in response to agonists. However, it was more than 20 years before Michell proposed a hypothesis suggesting a functional link between this lipid turnover and intracellular calcium homeostasis that spurred a new wave of interest, cumulating with the description of IP_3 as a second messenger by Berridge, Irvine, and their colleagues. This was followed by the later discovery of phosphatidylinositol 3 kinase by Cantley and co-workers. These significant advances were, however, just the beginning of a field that now encompasses every conceivable area of cell biology. The all-pervasiveness of these compounds in cellular regulation is largely facilitated by the unique structure of inositol which, when phosphorylated either in its native or lipid-derived forms, leads to the production of more than 30 unique isomeric forms with multiple functions. The tools of analysis necessarily reflect that complexity. The field is now divided into three major areas. These are the role of IP_3 in handling intracellular calcium, the inositol lipids (and particularly the 3-phosphorylated lipids), and the higher phosphorylated inositol polyphosphates. Within the last 20 years, there have been two excellent methodological books on inositol compounds, edited by Irvine and Shears, respectively, each separated by approximately a decade. This new inositide methods book not only introduces the basic methodological tools to measure inositol lipids and phosphates but also reflects new approaches that have become available in the last 10 years, including RNA silencing and the use of fluorescently labeled PH domains to measure inositides in real time in live cells as well as new sensitive methods to measure mass of both phosphates and lipids. Inositol pyrophosphates are an important current area of inositol phosphate research, and this is reflected in a number of chapters within the book. After an overview of methodologies, we start with inositol phosphates, move on to the IP_3 receptor and PLC, and then focus on the inositol lipids in the last section. Thus, the aim of this book is to compile many of the techniques that underscore phosphorylated inositol cell biology under one roof.

Stockholm, Sweden   *Christopher J. Barker*

Contents

Contributors

HIDEAKI ANDO • *Oregon Stem Cell Center, Oregon Health & Sciences University, Portland, OR, USA*

CRISTINA AZEVEDO • *Medical Research Council (MRC) Cell Biology Unit and Laboratory for Molecular Cell Biology, Departments of Cell and Developmental Biology, University College London, London, UK*

CHRISTOPHER J. BARKER • *The Rolf Luft Research Center for Diabetes and Endocrinology, Karolinska Institutet, Stockholm, Sweden*

MATTHEW BENNETT • *Medical Research Council (MRC) Cell Biology Unit and Laboratory for Molecular Cell Biology, Departments of Cell and Developmental Biology, University College London, London, UK*

PER-OLOF BERGGREN • *The Rolf Luft Research Center for Diabetes and Endocrinology, Karolinska Institutet, Stockholm, Sweden*

RASHNA BHANDARI • *Laboratory of Cell Signaling, Centre for DNA Fingerprinting and Diagnostics (CDFD), Nampally, Hyderabad, India*

ADAM BURTON • *Medical Research Council (MRC) Cell Biology Unit and Laboratory for Molecular Cell Biology, Departments of Cell and Developmental Biology, University College London, London, UK*

LUCIO COCCO • *Cellular Signaling Laboratory, Department of Anatomical Sciences, University of Bologna, Bologna, Italy*

FRANK T. COOKE • *Department of Structural and Molecular Biology, University College London, London, UK*

NULLIN DIVECHA • *CRUK Inositide Laboratory, Paterson Institute for Cancer Research, Manchester, UK*

IRENE FAENZA • *Cellular Signaling Laboratory, Department of Anatomical Sciences, University of Bologna, Bologna, Italy*

ROBERTA FIUME • *Cellular Signaling Laboratory, Department of Anatomical Sciences, University of Bologna, Bologna, Italy*

OLOF IDEVALL-HAGREN • *Department of Medical Cell Biology, Uppsala University, Uppsala, Sweden*

CHIHIRO HISATSUNE • *Laboratory for Developmental Neurobiology, Brain Science Institute, RIKEN, Saitama, Japan*

CHRISTOPHER ILLIES • *The Rolf Luft Research Center for Diabetes end Endocrinology, Karolinska Institutet, Stockholm, Sweden*

ROBIN F. IRVINE • *Department of Pharmacology, University of Cambridge, Cambridge, UK*

KASTUHIRO KAWAAI • *Laboratory for Developmental Neurobiology, Brain Science Institute, RIKEN, Saitama, Japan; Department of Chemistry and Biological Science, Aoyama-Gakuin University, Kanagawa, Japan*

ANDREW J. LETCHER • *Department of Pharmacology, University of Cambridge, Cambridge, UK*

HONGYING LIN • *Institut für Biochemie und Molekularbiologie I: Zelluläre Signaltransduktion, Universitätsklinikum Hamburg-Eppendorf, Hamburg, Germany*

KARSTEN LINDNER • *Institut für Biochemie und Molekularbiologie I: Zelluläre Signaltransduktion, Universitätsklinikum Hamburg-Eppendorf, Hamburg, Germany*

GEORG W. MAYR • *Institut für Biochemie und Molekularbiologie I: Zelluläre Signaltransduktion, Universitätsklinikum Hamburg-Eppendorf, Hamburg, Germany*

TAKAYUKI MICHIKAWA • *Laboratory for Developmental Neurobiology, Brain Science Institute, RIKEN, Saitama, Japan; Calcium Oscillation Project, International Cooperative Research Project-Solution Oriented Research for Science and Technology, Japan Science and Technology Agency, Saitama, Japan*

KATSUHIKO MIKOSHIBA • *Laboratory for Developmental Neurobiology, Brain Science Institute, RIKEN, Saitama, Japan; Calcium Oscillation Project, International Cooperative Research Project-Solution Oriented Research for Science and Technology, Japan Science and Technology Agency, Saitama, Japan*

AKIHIRO MIZUTANI • *Laboratory for Developmental Neurobiology, Brain Science Institute, RIKEN, Saitama, Japan*

ANDREAS NAGEL • *The Rolf Luft Research Center for Diabetes end Endocrinology, Karolinska Institutet, Stockholm, Sweden*

SARA MARIA NANCY ONNEBO • *Medical Research Council (MRC) Cell Biology Unit and Laboratory for Molecular Cell Biology, Departments of Cell and Developmental Biology, University College London, London, UK*

JAMES C. OTTO • *Department of Pharmacology and Cancer Biology, Howard Hughes Medical Institute, Duke University Medical Center, Durham, NC, USA*

TREVOR R. PETTITT • *CRUK Institute for Cancer Studies, University of Birmingham, Birmingham, UK*

ADOLFO SAIARDI • *Medical Research Council (MRC) Cell Biology Unit and Laboratory for Molecular Cell Biology, Departments of Cell and Developmental Biology, University College London, London, UK*

MICHAEL J. SCHELL • *Department of Pharmacology, Uniformed Services University of the Health Sciences, Bethesda, MD, USA*

TRACI SPEED • *Department of Pharmacology and Molecular Sciences, Johns Hopkins University School of Medicine, Baltimore, MD, USA*

ANDERS TENGHOLM • *Department of Medical Cell Biology, Uppsala University, Uppsala, Sweden*

GABRIELLA TETI • *Department of Anatomical Sciences, University of Bologna, Bologna, Italy*

J. KENT WERNER, JR. • *The Solomon H. Snyder Department of Neuroscience, Johns Hopkins University School of Medicine, Baltimore, MD, USA*

ANNE WUTTKE • *Department of Medical Cell Biology, Uppsala University, Uppsala, Sweden*

JOHN D. YORK • *Department of Pharmacology and Cancer Biology, Howard Hughes Medical Institute, Duke University Medical Center, Durham, NC, USA*

Chapter 1

The Role of Inositol and the Principles of Labelling, Extraction, and Analysis of Inositides in Mammalian Cells

Christopher J. Barker and Per-Olof Berggren

Abstract

Inositides have an important impact on diverse areas of cellular regulation. However, since this area has grown exponentially from the mid 1980s onwards, many workers find themselves relatively new to the field. In this chapter, we establish a broad foundation for the rest of the book by covering some important principles of inositide methodologies. The focus is especially directed to those methods or aspects of methodology not covered in detail in other chapters. This includes the often neglected influence of the inositide precursor, inositol, and important background information relating to the labelling and extraction of inositides from cells and tissues. This introductory section also gives a "birds eye" view of important methods and protocols found within this volume and hopefully acts as a touchstone to assess which of the methodologies described within this book is most appropriate for your particular study(ies) of inositides.

Key words: *myo*-inositol, D-*myo*-inositol 3-phosphate synthase, Inositol phosphate, Inositol lipid, Equilibrium labelling, Acid extraction

1. Introduction

In this overview chapter, we discuss some of the underlying principles of the methods employed by inositide researchers to study inositol phosphates and lipids. This discussion thus serves as an extended introduction to the book as a whole and highlights various assumptions made by workers in the field. This background information is important as it gives both existing researchers, and those wanting to enter the field, a solid foundation. In particular, this chapter focuses on areas not covered in detail elsewhere in the book with a special emphasis on the role of *myo*-inositol itself, the precursor of all inositol phosphates and lipids.

Christopher J. Barker (ed.), *Inositol Phosphates and Lipids: Methods and Protocols*, Methods in Molecular Biology, vol. 645,
DOI 10.1007/978-1-60327-175-2_1, © Humana press, a part of Springer Science+Business Media, LLC 2010

2. Inositol

In all the studies on inositol phosphates and lipids, inositol itself is probably the most neglected variable, despite the fact that all inositides have it as their backbone (Fig. 1). From a methodological standpoint, people appreciate the role of inositol concentration when considered in the context of labelling cells with [³H]*myo*-inositol (which we cover in the next section). This appreciation stems from the fact that most labelling protocols use reduced inositol concentrations to maximize labelling and thus sensitivity.

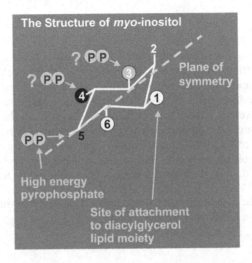

Fig. 1. Structure of *myo*-inositol. The figure shows the structure of *myo*-inositol with D-numbering nomenclature. The D-1 position relates to the site of attachment of the glycerol moiety of inositol lipids. At each position is a H– and OH– group, but only the OH groups are shown, represented by projected angled lines from the carbon atoms. The hydroxyl groups are modified with phosphate groups to form various inositides. Because inositol is a meso-compound with a plane of symmetry between the D-2 and D-5 positions, substitution by phosphate groups in the D-1/3 or D-4/6 positions leads to the generation of enantiomers giving the possibility of generating 63 different inositol phosphate motifs – and this does not include pyro-phosphorylation. Some confusion exists, not least among suppliers of inositol phosphates, because there was a change in nomenclature from the D or L form to just the D form. Essentially the difference between the two numbering systems is that in the D form the 1 position is defined by the site of attachment of inositol to glycerol in phosphatidylinositol (see Fig. 1). The carbons are then numbered sequentially in an anti-clockwise manner. In the L-nomenclature the L-1 position is the D-3 position and the labelling goes clockwise. The practical upshot of this is that the L-Ins(1,4,5,6)P_4 described in (20) is *exactly* the same molecule as D-Ins(3,4,5,6)P_4, whereas D-Ins(1,4,5,6)P_4 is the enantiomer of D-Ins(3,4,5,6)P_4, inseparable by conventional HPLC and requiring enantiomer specific degradative analysis to assign the complete structure. At one time, suppliers were selling Ins(1,4,5,6)P_4 without assignment and it was, in fact D-Ins(3,4,5,6)P_4! There is also some discussion about the position of the second pyrophosphate groups in InsP_8. A recent publication [63] and the Mayr chapter in this book (Chapter 7) suggest that in mammalian InsP_8 or Bis-diphosphoinositol tetrakisphosphate, the positions are 5- and 1/3 and not 5- and 4/6 as previously suggested (64).

However, in most studies, no deliberate attempt has been made to align the inositol concentrations in the culture media with that of the serum of the animal from which the cell is originally derived. There are some necessary practical and economic reasons why you might not try to mimic true inositol concentrations. Nonetheless, we now highlight some issues which, while they may not be avoided, should at least be in the mind of the investigator when doing experiments on inositol lipids and phosphates.

Before we go any further, the question may be asked: is inositol concentration really so critical? An example of its relevance is the late recognition of the inositol phosphate inositol 1,2,3-trisphosphate ($Ins(1,2,3)P_3$) as a common constituent mammalian cells (1, 2). $Ins(1,2,3)P_3$'s cellular concentrations are proportional to the prevailing inositol concentrations (2), and in the few close examinations of inositol phosphate structures that had been carried out, low concentrations of inositol were used to maximize labelling of the less abundant inositol phosphates (e.g. (3)). Recent work in yeast (*Saccharomyces cerevisiae*), however, has demonstrated that changes in inositol concentration can elicit the expression of more than 100 genes (4–6), a sobering thought. We have been lulled into a sense of false security by the fact that in mammalian cells many inositol derivatives are fairly resistant to changes in extracellular inositol.

When we consider the role of inositol in studies of inositol lipids and phosphates, a good starting point is the normal plasma inositol concentrations. Here, an immediate distinction must be made between adult and foetal serum. The range of inositol concentrations for adult mammals whose cells are used in experimental work is 30–100 µM, human being around 30 (e.g. (7, 8)), rat 63, rabbit about 40 (8, 9), and guinea pig 94 µM (9). In contrast, foetal serum is often more than an order of magnitude higher in all mammalian species examined (8, 10, 11). For example, foetal calf serum has been estimated to be around 1 mM (10, 12) compared to 50 µM for calf (12) or adult bovine serum (9, 11). At least in humans, higher levels of inositol persist up to a year following birth (e.g. see (13)). When considering serum inositol concentrations, especially relating to adult animals, the variability of the reported concentrations in a given animal is probably most often related to difference in diet (14, 15) rather than the methods of measurement. Another important consideration regarding inositol concentration is that not all cells are bathed by the normal circulating concentrations of inositol in plasma. For example, cerebral spinal fluid (CSF) in humans (7) and other mammals (16) has an inositol concentration four times higher than plasma.

There are two immediate consequences arising from the differences between adult and foetal serum inositol concentrations. Firstly, it is important to know whether the cell line was derived from foetal or adult tissues and how long ago. Prolonged culture is likely to lead to the cell's adaption to the prevailing inositol concentration, which may or may not reflect physiology. Secondly, many cells are grown in 5–10% foetal calf serum, which could contribute at least 100 μM inositol to the culture conditions, dependent on the actual concentration of inositol in the serum (10–12). Given that most media have an inositol concentration below 40 μM, it is the contribution of the serum inositol that is likely to be the deciding factor in determining the final inositol concentration in the medium. One notable exception is RPMI 1640 media whose inositol concentration is 194 μM. The high concentration of inositol in RPMI 1640 media is actually important for the medium to function as it has half the physiological Mg^{2+} concentration, which will lead to inefficient inositol uptake (17). Therefore, the high inositol overcomes a constraint on inositol transport, normalizing inositol supply.

The paragraph above discusses the physiological concentration of inositol. However, in most cases, the critical issue is what is the minimum inositol concentration one can use to maintain normal cell growth? Harry Eagle and colleagues defined a minimum essential medium (MEM) in which a variety of human and a few mouse cell lines would grow (18, 19). Although the minimum concentration of inositol that all human cell lines studied would grow optimally was 1–3 μM, this may be insufficient to maintain growth of mammalian cells in general. This is because normal plasma inositol concentrations are higher in other mammals than they are in humans ((7–9), see above). Interestingly, in Eagle's MEM (EMEM), the minimum essential medium that has been given his name, the inositol concentration is actually set at the higher concentration of 11 μM. This gives a margin for error, making it potentially suitable for the growth of all adult mammalian cells. Finally, with a potentially variable contribution of serum inositol for both experiments related to cell growth and those in which inositol labelling is carried out, inositol can be dialysed out of serum so that the inositol concentration can be accurately controlled. It has been suggested that dialysis can cause loss of important factors required for growth. This is certainly the case when using standard dialysis tubing used for protein dialysis. However, we have found that using lower molecular cut off (1,000 MW, e.g. from SpectraPor) has given us no restriction of growth with a variety of different cell lines. One important factor that could be lost in serum dialysis and may be important for long-term cell culture, rather than just labelling, is selenium, but this can be added back to the medium if necessary. Of course, it is always necessary to test serum so prepared in your own experimental system.

So far, we have avoided discussing the main reason why people do not use physiological inositol concentrations: it is simply a question of economics. The higher the inositol concentration, the more [^{3}H]*myo*-inositol is needed to detect the inositide, and no lab has infinite reserve of cash to spend on inositol.

We have adopted using 50 µM as a standard concentration with pancreatic beta cells and we would recommend testing this at least once against lower concentrations in experimental protocols.

Another issue regarding inositol is that in some experimental protocols, especially those involving agonist stimulation, cells are transferred from a medium that probably contains nominally about 100 µM inositol to an essentially "inositol-free" balanced salt solution. Under these conditions, there is constant efflux of inositol (C.J. Barker, unpublished observations). Therefore, care must be taken to check whether this efflux affects cellular readouts. That is, does the presence or absence of extracellular inositol influence the measured cellular responses? Related to this, it is important if you continue to supply inositol during a period of stimulation that it is of the same specific activity as that which the cells have been labelled in. Adding "cold" inositol at this stage will lead to a different kind of experiment, effectively a cold chase, the results of which may be interesting, but not intended.

Last, but by no means least, a final concern regarding inositol itself is: where is it coming from? Figure 2 indicates that inositol is either synthesized *de novo* from glucose or transported from the extracellular medium via specific passive or active transporters. This is an important point which we will discuss more fully in the section on labelling with [³H]*myo*-inositol, as it is in this scenario that the inositol source becomes the most critical.

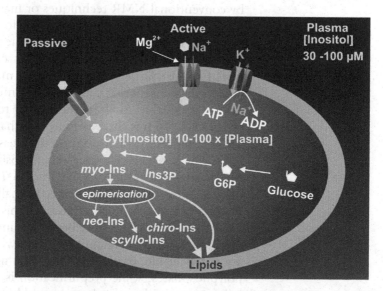

Fig. 2. Sources and metabolism of inositol in cells. The figure depicts the two sources of *myo*-inositol in mammalian cells, either *de novo* synthesis via inositol synthase, or by active uptake by sodium-dependent *myo*-inositol active transporters (SMITS). *Myo*-inositol can be further epimerized into *neo*-, *scyllo*-, and *chiro*-inositol. *Chiro*-inositol forms part of the structure of inositol-glycan linked lipids.

3. How Do We Measure Inositides?

Historically, two approaches to study inositides have been used: radioisotopic labelling with precursors or more direct mass measurements. Looking back at the literature, labelling cells or tissues with either $[^3H]$*myo*-inositol or $[^{32}P]$Pi have overshadowed mass-based techniques, but recently mass-based techniques are becoming more prevalent (see Chapters, 4, 7, and 13). However, the biggest growth area has been the use of fluorescently tagged reporters, especially in the measurement of inositol lipids (see Chapter 14 and references therein). By reporter, we mean an engineered construct, often a PH (Plekstrin homology) domain, linked to a fluorescent tag (usually GFP or its enhanced form, EGFP). An individual PH domain specifically recognizes a given inositide isomer, usually a lipid. This technique allows the study of inositide metabolism in single cells and in real time, thus providing both spatial and temporal information.

Both mass measurements and reporter constructs are presented within the following individual chapters of this volume. Until now, the main restrictions for using mass techniques relate to the amount of material required and the ability to distinguish inositol compounds from other cellular constituents and from each other. The later comment refers to the fact that rigorous identification of a given inositide requires both its complete separation from other related isomers and the determination of its enantiomeric structure. Stereo-specific data cannot be achieved by conventional NMR techniques or mass techniques and therefore they do not ascribe enantiomeric forms, e.g. D-Ins(3,4,5,6)P_4/Ins(1,4,5,6)P_4 (3, 20). A technique that could discriminate between enantiomeric forms is X-ray crystallography, but this would require large amounts of pure material to start with. Of course, mass assays that can selectively measure individual inositides are invaluable. Such methods exist to measure inositol lipids (e.g. (21–23)). It is also worth noting that there is a commercially available binding protein assay for the measurement of Ins(1,4,5)P_3 (24), which is selective over other inositol phosphates and that can be adapted to measure PtdIns(4,5)P_2. This assay is supplied by a division of GE Healthcare, formally known as Amersham Pharmacia Biotech. Furthermore, in this volume a new, sensitive mass assay for InsP_6 is described by Irvine's group (Chapter 4).

However, in the majority of cases where such selective assays are not available, particularly the mass measurements of inositol polyphosphates, some prepurification is needed. This is particularly an issue with inositol mono- and bisphosphates as most techniques to separate Ins$P$s rely on ion exchange and many sugars, existing at considerably higher concentrations than inositol phosphates, possess one or two monoester phosphate groups.

Fortunately, there are techniques to at least remove nucleotide contaminants (25). Moreover, for reasons that will become clearer in the next section, direct mass measurements can avoid some complications in the interpretation of data that are introduced when using radioisotopic techniques.

3.1. Principles of Labelling with [³H]myo-Inositol

At the outset it is important to be clear that the cell or tissue under study is permeable to inositol as some cells, e.g. human erythrocytes, cannot efficiently take up inositol. [³H] (normally added in the 2-position, [2-³H] or less common [1,2-³H]) is the label of choice in most studies of inositol phosphates and some lipid studies. An alternative is [U-¹⁴C], but the 10 to 20-fold greater expense of [¹⁴C]myo-inositol versus [³H]myo-inositol usually discounts its use in routine experimental settings. It is, however, very useful when wanting to make a direct comparison between steady-state levels of inositols versus their turnover rate. In this situation, cells are labelled to isotopic equilibrium with [¹⁴C]myo-inositol and then "pulsed" with high amounts of [³H]myo-inositol. This was important to, for example, delineate the source of $Ins(3,4,5,6)P_4$ in mammalian cells (26). In this context, the reverse protocol would be prohibitively too expensive. A less elegant alternative is to label matched dishes of cells to either equilibrium or for a short time (e.g. 4 h) and do a ratio between them (27). There are a number of factors that need to be considered when using [³H] as a radio label. First, concerning the label itself, secondly related to the side effects of using the label and thirdly, the fate of the label in the cell or tissue being studied.

Because of the nature of radioactivity, it is inevitable that there is a slow but steady build up of radiolysis products that arise from the radiation-mediated degradation of the labelled molecule. These degradation products are of importance, as several are polar and have elution positions on HPLC similar to *bone fide* inositol polyphosphates. A clear example of this kind of contamination is shown in a paper by Staddon et al. (Fig. 1 in that paper) (28). Figure 3 in this chapter shows [³H] inositol with or without pretreatment down a simple anion-exchange cartridge. It can be seen that peaks approximately corresponding to GroPInsP and GroPInsP$_2$ disappear when the inositol is "cleaned up." Publications purporting to study these glycerolphosphoinositol derivatives need to be examined carefully to ensure the GroPIns species were identified by more than co-elution on an HPLC column. The early recognition of contaminants in the InsP$_3$ region led to the inclusion of small "pellets" of undefined absorbent material (possibly ion-exchange media) to "mop up" the radiolysis products as they build up. Inositol supplied by GE Healthcare (originally Amersham), for example, can be ordered with such a "cleaning" agent. (*Unfortunately, from 2010 GE Healthcare will stop producing [³H]inositol completely, it continues to be available from other*

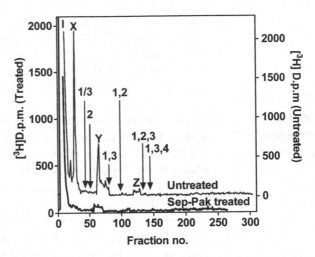

Fig. 3. Removal of radiolysis products from commercial [³H]*myo*-inositol. [³H]*myo*-inositol (3 μCi) obtained from American Radiochemicals Inc. was either untreated or passed through a Sep-Pak QMA anion-exchange cartridge. The resulting samples were injected onto a 25-cm Whatman Partisphere SAX HPLC column and a standard elution protocol based on ammonium phosphate was employed (2). [¹⁴C]inositol phosphate standards were added into the non-filtered preparation prior to injection. They are marked, rather than plotted, to make the figure clearer. Fractions were collected and radioactivity determined using a Packard Liquid scintillation counter using a dual label protocol. For the sake of clarity, Ins*Pn* isomers are identified by numbers representing the position of phosphate groups on the inositol ring. A "/" indicates when the Ins*P* exists as an enantiomeric form.

suppliers, e.g. ARC). One note of caution here is that even when such "cleaned" inositol is used, these radiolysis products of inositol build up during the long-labelling protocols required to label higher inositol polyphosphates to equilibrium. Of more concern is that in pulse-labelling experiments, in which a large amount of inositol is given to the cells for short periods of time, say up to 4 h (e.g. (29)), these inositol radiolysis products severely contaminate the so-called Gro*P*Ins*P* and Gro*P*Ins*P*₂ peaks corresponding to the deacylated products of PtdIns and PtdIns4P (29). Moreover, even in equilibrium labelled cells, these can contribute to up to 30% of the Gro*P*Ins peak (C.J. Barker and D.J Candy, unpublished observations). Fortunately, these inositol radiolysis products are not normally an issue with extracted inositol lipids.

The second consideration when using labelled precursors is the effect of the isotope itself. Since it is now understood that inositol lipids and phosphates are universally distributed in cells and particularly the nucleus (30), radiation damage will be a clear, unwanted side effect of using radiolabelled precursors (31). Of course, dose-dependency is critical here, but experience shows that different cell types are markedly different in respect to their radiation sensitivity. A ready indicator of the susceptibility of the cells to radiation damage is to measure the growth rate of the cells with and without label. Such an approach revealed that HL60 cells, a haematopoietic cell line, suffered growth inhibition at [³H] inositol concentrations greater than 2 μCi/ml. WRK-1 cells, a rat mammary tumour cell line, were more robust and could be routinely grown in 10 μCi/ml [³H]*myo*-inositol (e.g. (32)). However, prolonged incubations of these cells, longer than 7 days, revealed growth inhibition unless the concentration was reduced to 5 μCi/ml (29). The authors suspect that many investigators have not carefully checked this parameter in their cells of interest. The use of [³²P] is clearly a concern here, given that it emits beta particles with much higher energy than both [³H] and [¹⁴C]. However, since [³²P]Pi reaches isotopic equilibrium with the cellular ATP pool, and thus the phosphorylated inositols, in an hour or so, rather than days (see Chapter 12), the damage may not be as serious as one might expect as the time period of exposure is much shorter.

The third consideration is that fate of the labelled inositol in the cell or animal system. The assumption is that the labelled *myo*-inositol molecule remains unchanged. Certainly, in most cell types, with the exception of kidney cells, the inositol ring is thought to be metabolically stable. However, because of the presence of *myo*-inositol oxygenase (MIOX) in the kidney (33), whole animal labelling is fraught with difficulty in interpretation. For example, studies have indicated the transformation of [³H]*myo*-inositol-label into ATP (34). Moreover, using more stringent techniques, recent work has indicated at least the presence of MIOX mRNA in sciatic nerves, liver, heart, and lungs (35). A more important issue in terms of the transformation of the labelled *myo*-inositol is the fact that many mammalian cells express the enzymatic machinery that epimerizes *myo*-inositol to other forms of inositol such as *scyllo*-, *neo*-, and *chiro*-inositol (36–39) (Fig. 2). However, in the case of the epimerization of *myo* to *scyllo*-inositol, [2-³H]-labelled *myo*-inositol would effectively loose the label and not contribute to any further analyses. So far, *chiro*-inositol has only been associated with proteoglycan structures and the water soluble derivatives based on these molecules, which may serve as insulin mediators. *Chiro*-inositol has not been demonstrated to form inositol phosphate structures (39), at least in mammalian cells.

3.1.1. Equilibrium Labelling with [³H] myo-Inositol

In some experimental approaches, it is important to label for only a brief period of time to look at turnover of inositol phosphates or lipids. This has led to the characterization of important metabolic relationships (e.g. (26, 27)). Also, the fact that the steady-state level of an inositide does not change may mask a higher degree of turnover critical for its function, as in, for example, inositol pyrophosphates (40–42). While it is often useful to label for only brief periods of time when one wants to study turnover of inositides, in order to gain information on mass changes using labelling protocols, the principle to follow is that the material is labelled to isotopic equilibrium. That is, the cells are labelled until incorporation of label in an individual inositide reaches a steady state where there is no additional incorporation and all pools of inositides are assumed to have come to isotopic equilibrium with the labelled inositol. It is important to bear in mind that equilibrium is reached faster by some inositides than others. Inositol lipids usually reach equilibrium between 24 and 48 h, whereas highly phosphorylated inositols and especially putative degradation products of inositol hexakisphosphate can take 120 h or more (e.g. (29)). This equilibrium labelling approach is subject to several caveats, sometimes ignored. The two principal ones are that the cell type concerned cannot make inositol from glucose, i.e., it does not possess an active inositol synthase and that the cells are dividing and can undergo several division cycles during labelling. We will tackle the first of these limitations in more depth.

3.1.2. Can Inositol Be Derived from Another Source (i.e., Glucose) in My Cell Type?

In yeast, the issue of labelled inositol being diluted with *de novo* synthesis has been effectively dealt with by researchers using yeast inositol auxotrophs in which inositol synthase is naturally absent (*S. Pombe*) (43) or has been deleted (e.g. various *S. cerevisiae* mutants, see ref. 44). However, the same cannot be said of mammalian cells. One rule of thumb that is not, unfortunately, 100% reliable is that mammalian cells that have an active inositol transporter (largely of the SMIT family (45)), by and large, do not synthesize their own inositol, and cells that *do not* possess active transporters synthesize inositol *de novo*, e.g. testicular cells (46). It must be remembered that when inositol synthase has been reported in a given tissue, it is likely that there is heterogeneity to its distribution in the individual cell types. For example, crude extracts of brain will reveal both inositol synthase and an active inositol transporter to be present; however, a closer examination has revealed that it is likely that neurones cannot synthesize inositol de novo (47), whereas inositol synthase can be found within the surrounding vasculature (48). One exception to this rule may be the retinal cells, which apparently possess both active inositol transporters and inositol synthase (49–51).

Following the yeast model in which inositol synthase is inducible at low environmental inositol (52), it might be expected that growing cells in low inositol, during, for example, attempts to maximize labelling, would activate synthase activity and distort the relationship between equilibrium labelling and mass. This concern can be addressed experimentally by attempting to grow cells in low inositol conditions while labelling with [U-^{14}C] glucose. The lipids are then extracted and phosphatidylinositol degraded into its individual components, inositol, glycerol, and fatty acids, using methods described in this book. If inositol is being made *de novo*, the resulting inositol ring should be [^{14}C] labelled. A cruder but effective method would be to boil the extract in a sealed tube in the presence of 1 M HCl for 24–48 h. Under these conditions, the only survivors are inositol and other polyols. Separation of these remnants on a carbohydrate HPLC column (e.g. (2)) with a [^{3}H]inositol internal standard will indicate whether inositol became [^{14}C]-labelled and hence whether inositol synthase is a player or not. However, recent studies have indicated that mammalian inositol synthases, at least at the mRNA level, are not inducible by inositol starvation in the same manner as in yeast and other free-living single-cell organisms (53). An important caveat is that increased glucose or even lithium can change the mRNA levels of the inositol synthase in mammalian cells (53, 54).

Returning to Eagle's original studies, it is clear that although only fibroblasts were able to grow indefinitely without added inositol (19), and thus had an active inositol synthase, he and his colleagues noted that several cell lines actually also made inositol from [^{14}C] glucose even though this synthesis was not enough to sustain cell growth (55). A notable, commonly used human cell line that was able to synthesis inositol was the HeLa cell, still used by inositide experimentalists today. Some cells used in culture are even completely independent of extracellular inositol, e.g. CHO cells (56).

Equilibrium labelling also gives the opportunity to gain a crude estimate of the cellular inositol phosphate concentration. If one grows cells from a low density (say 1/10 confluence) in [^{3}H]*myo*-inositol with a known specific activity, it is possible to make a crude estimate of the inositol phosphate concentration. For this an estimate of cell volume is also required. For cells in suspension, the cell diameter can be measured microscopically, but for adherent cells, this is more problematic. However, after trypsinization, many adherent cells assume a spherical form. The data obtained is, of course, a cell-wide average of the concentration of the particular inositol phosphate isomer, but at least it gives a ball park estimate for what the actual cellular concentration might be. This approach has been used to estimate inositol phosphate concentration in both adherent and suspension cells (2, 29, 57, 58).

A final note of caution comes from studies done while CJB, one of the authors of this chapter, was a research fellow in the Kirk/Michell laboratory. Here, a rat mammary tumour cell line (WRK-1), which was not able to make inositol de novo (R. Fox, C.J. Barker, R.H. Michell, and C.J. Kirk, unpublished observations), nonetheless exhibited anomalous labelling patterns, see Fig. 4. To summarize this work, equilibrium labelling of growing or quiescent WRK-1 cells indicated that both lipids (Fig. 4a) and phosphates (Fig. 4b) achieved steady-state equilibrium.

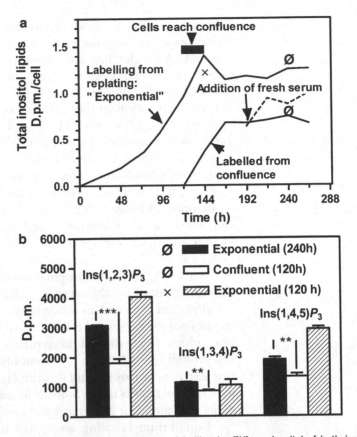

Fig. 4. *Discrepancies in the equilibrium labelling by* [³H]*myo-inositol of both inositol lipids and phosphates of WRK-1 cells*. The figure illustrates that more [³H]*myo*-inositol is incorporated into the inositides from cells labelled during exponential growth than those cells labelled following growth arrest and "confluence," even though in both cases a steady-state isotopic equilibrium has been achieved. WRK-1 cells at a tenth confluent density were replated into 3.5 cm dishes either in the presence or in the absence of 5 µCi/ml of [³H]*myo*-inositol. When the cells reached "confluence" following 120 h growth 5 µCi/ml of [³H]*myo*-inositol was added to the previously unlabelled dishes. At the times indicated (X or Ø), inositol phosphates and lipids were extracted using a TCA/ ether protocol and inositol phosphates separated on HPLC, exactly as described in ref. 28. The "Ø" symbol denotes samples taken from identical times post-confluence either from cells labelled throughout exponential growth and the subsequent confluent state (240 h) or just labelled following confluence (120 h). The cells are thus at an equivalent stage in the growth cycle. (a) Total inositol lipids labelled under the conditions described above and with the addition of fresh serum at 192 h (*dashed line*) to dishes labelled from confluence. (b) InsP_3s from extracts of cells labelled from replating under "exponential" conditions after 120 or 240 h or labelled from "confluence" for 120 h. A direct comparison can made between the amount of labelled InsP_3 recorded in cells at identical growth stages ("Ø") but labelled to equilibrium during either exponential growth or following growth arrest and confluence. Comparison between cells labelled for the same time, 120 h (X vs. Ø) is difficult to interpret as the cells may be phenotypically distinct, being at different stages of the growth cycle. **$P > 0.01$, ***$P > 0.005$ using paired *t*-test. The experiment was carried out in triplicate, two other experiments gave similar results.

However, the final level of label was higher in the cells labelled during the growth phase. Because there are changes in inositides after the cells becoming confluent, stop growing and enter a "G_0" - like state (29), the most valid comparison to be made is between the cells 240 h after the original plating. Addition of fresh serum promotes further incorporation of label into lipids (Fig. 4a), again suggesting that the phenomenon is cell cycle related. These data suggest that there are pools of inositol lipids and phosphates that are metabolically inactive, or compartmentalized, and only become active during cell growth. One inositol polyphosphate that fits this notion is inositol 1,2,3-trisphosphate as it labels slowly to isotopic equilibrium, but has a very high turnover rate during cell cycle progression (29). However, it should be noted that $Ins(1,4,5)P_3$ also exhibits this anomalous labelling. Since there is a similar anomaly in the inositol lipids, the $Ins(1,4,5)P_3$ is believed to be derived from – that is total lipids (Fig. 4a), but more critically $PtdIns(4,5)P_2$ (data not shown) – it is likely this phenomenon could affect classic signal transduction pathways. Indeed, discrepancies between the total cellular mass of $Ins(1,4,5)P_3$ and the actual concentration required to elicit mobilization of Ca^{2+} from intracellular stores have been noted by Putney's group (59, 60) among others and may reflect a similar or related phenomenon. These findings have a direct relevance in the labelling of primary cells, which either have a low rate of cell division or are in fact terminally differentiated. Under these conditions, isotopic equilibrium will be achieved, but if stimulation by, for example, growth factors leads to entry into the cell cycle, the relationship between amount of label and mass will break down.

This last section underscores the methodological weaknesses of labelling by [^{3}H] *myo*-inositol and the need for the measurement of mass directly. We have found it valuable to check out labelling studies against mass techniques in order to get a clearer picture of what is going on.

3.2. Extraction Techniques

Following the labelling of cells, the quantitative extraction of both lipids and phosphates is desired. Inspection of the appropriate protocols in several of the chapters in this book will give the reader an appreciation of the major methods employed. In general, the extraction techniques usually involve an acid-based kill and separation into aqueous- and organic-soluble fractions. When inositol phosphates are extracted, a standard complementary lipid extraction protocol that is used is based on a modified acidified Bligh Dyer technique ((61) and see Notes of Chapter 2). However, as is pointed out elsewhere in this book (Chapter 12), more specialized extraction techniques are required if you wish to quantitatively recover the minor $PtdInsP_2$s and $PtdInsP_3$ that are the products of PI3 kinase. Unfortunately, these extraction techniques are not then optimized for inositol polyphosphate extraction. So it may not be possible to efficiently extract both lipids and phosphates at the same time. Since lipid extraction is covered in a later chapter (Chapter 12), we will focus here on inositol phosphates.

3.2.1. Inositol Phosphates

Historically, two principal acid extraction techniques have been used and are trichloroacetic or perchloric acid based. The necessary neutralization of these acid extracts is then achieved by ether washing or perchlorate precipitation. In our hands we have preferred to use trichloroacetic acid as we find that it gives a "cleaner" preparation when separating the neutralized inositol phosphate extract by HPLC, especially when extracting inositol phosphates from larger dishes of cells (e.g. 9–10 cm), or in other words when more cellular material is involved (3). In contrast, when using simple ion-exchange columns to separate different inositol phosphate classes, i.e. mono-, bis-, tris-, and tetrakisphosphates, the percholate precipitation is more useful. The main issue with acid extraction is that acid-labile inositol polyphosphates are not quantitatively recovered or may be completely lost. This is true of cyclic inositol phosphates, but it is also to some extent true for inositol pyrophosphates, although these, in our hands, can be preserved by rapid neutralization using the TCA/ether wash technique and an appropriate buffer cocktail (see Chapter 2). A phenol-based extraction, which does not allow the quantitative extraction of inositol lipids, has been used to quantitatively extract the acid-labile cyclic inositol phosphates (e.g. (2, 3, 62)). We have found that the quantitative recovery of any inositol polyphosphate more phosphorylated than a trisphosphate is not possible when dealing with small amounts of cellular material (3.5 cm dishes), without the inclusion of a cold carrier. In our experiments, we add a relatively crude phytic acid preparation to the TCA at a concentration of 250 μg/mL, which helps to preserve the more phosphorylated inositols. The inclusion of cold carrier in an extraction procedure should be avoided if the samples are subsequently going to be later used to perform structural determinations using chemical-based degradation and the assignment of structure based on resulting polyol forms (e.g. (2, 3)). When studying inositol polyphosphates, it is worthwhile doing a neutral extraction at least once and comparing to the standard acid extraction used. The reason for this is illustrated in Fig. 5a, b. Inositol phosphates from pancreatic beta cells were extracted with either the TCA extraction technique or the neutral phenol technique. What is immediately apparent is the dramatic reduction of the Ins1/3P and Ins2P peaks in the monophosphate region of the chromatogram when comparing neutrally extracted cells (Fig. 5b) with conventional acid extraction (Fig. 5a). This reduction in the size of the Ins1/3P and Ins2P peaks would normally be attributed to the preservation of cIns$P$, which runs slightly before the non-

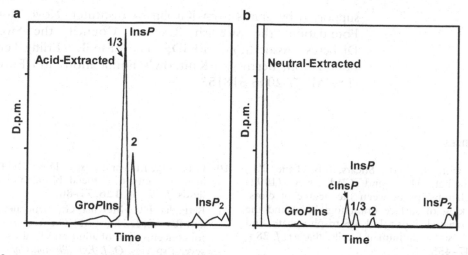

Fig. 5. Comparison between acid and neutral extraction of pancreatic beta cells. Insulin-secreting HIT T15 cells were either acid extracted using TCA (5% w/v) (**a**) or neutrally extracted using a phenol/chloroform saturated phosphate buffer (62). Extracts were then separated using conventional anion-exchange HPLC (see Fig. 3). The region of the chromatogram between inositol and inositol bisphosphates is depicted. For the sake of clarity, Ins*P* isomers are identified by numbers representing the position of phosphate groups on the inositol ring. A "/" indicates when the Ins*P* exists as an enantiomeric form.

cyclized forms (see Fig. 5b) and whose acid-mediated breakdown is thought to contribute to the Ins1/3*P* and Ins2*P* peaks. However, much more material has disappeared than can be accounted for from cIns*P*. We suggest the Ins1/3*P* and Ins2*P* are derived from acid-labile lipid glycan structures although this has not been demonstrated experimentally. What it does mean is the possibility that most of the Ins1/3*P* and Ins2*P* monophosphate peaks in HPLC profiles of [³H]-*myo*-inositol-labelled, acid-extracted cells are not derived from cIns*P* degradation and are in fact not representative of cytosolic inositol polyphosphates at all (see comments later in Chapter 2). Of course, the confusion is also compounded by the presence of inositol radiolysis products in this region too (see Subheading 3.1). Furthermore, it is worth noting that the *so-called* GroPIns peak in acid-extracted cells is much broader and larger than the same peak coming from neutrally extracted cells, suggesting other contaminants introduced by acid extraction.

Acknowledgments

The authors would like to thank Prof. R.H. Michell and Dr. C.J. Kirk for permission to use unpublished data obtained while C.J.B was a research fellow in their laboratory and Prof. R.F. Irvine for reading and commenting on the manuscript. This work was supported by grants from Karolinska Institutet, Novo Nordisk Foundation, the Swedish Research Council, the Swedish Diabetes Association, EFSD, The Family Erling-Persson Foundation, Berth von Kantzow's Foundation, and EuroDia (LSHM-CT-2006-518153).

References

1. McConnell, F.M., Shears, S.B., Lane, P.J., Scheibel, M.S., and Clark, E.A. (1992) Relationships between the degree of cross-linking of surface immunoglobulin and the associated inositol 1,4,5-trisphosphate and Ca^{2+} signals in human B cells. *Biochem J.* **284**, 447–455.

2. Barker, C. J., French, P. J., Moore, A. J., Nilsson, T., Berggren, P. O., Bunce, C. M., et al. (1995) Inositol 1,2,3-trisphosphate and inositol 1,2- and/or 2,3-bisphosphate are normal constituents of mammalian cells. *Biochem J.* **306**, 557–64.

3. Wong, N.S., Barker, C.J., Morris, A.J, Craxton, A., Kirk, C.J., and Michell R.H. (1992) The inositol phosphates in WRK1 rat mammary tumour cells. *Biochem J.* **286**, 459–468.

4. Santiago, T. C., and Mamoun, C. B. (2003) Genome expression analysis in yeast reveals novel transcriptional regulation by Inositol and Choline and new regulatory functions for Opi1p, Ino2p, and Ino4p. *J. Biol. Chem.* **278**, 38723–38730.

5. Jesch, S. A., Zhao, X., Wells, M. T., and Henry, S. A. (2005) Genome-wide analysis reveals Inositol, Not Choline, as the major effector of Ino2p-Ino4p and unfolded protein response target gene expression in Yeast. *J. Biol. Chem.* **280**, 9106–9118.

6. Gaspar, M.L., Aregullin, M.A., Jesch, S.A., and Henry, S.A. (2006) Inositol induces a profound alteration in the pattern and rate of synthesis and turnover of membrane lipids in *Saccharomyces cerevisiae. J Biol. Chem.* **281**, 22773–22785.

7. Servo, C., and Pitkänen, E. (1975) Variation in polyol levels in cerebrospinal fluid and serum in diabetic patients. *Diabetologia.* **11**, 575–580.

8. Baker, H., DeAngelis, B., and Frank. O. (1988) Vitamins and other metabolites in various sera commonly used for cell culturing. *Experientia.* **44**, 1007–1010.

9. Dawson, R.M., and Freinkel, N. (1961) The distribution of free mesoinositol in mammalian tissues, including some observations on the lactating rat. *Biochem J.* **78**, 606–610.

10. Campling, J.D., and Nixon, D.A. (1954) The inositol content of foetal blood and foetal fluids. *J. Physiol.* **126**, 71–80.

11. Battaglia, F.C., Meschia, G., Blenchner, J.N., and Barron, D.H. (1961) The free *myo*-inositol concentration of adult and fetal tissues of several species. *Q. J. Exp. Physiol. Cogn. Med. Sci.* **46**, 188–193.

12. Berry, G.T, Johanson, R.A., Prantner, J.E., States, B., and Yandrasitz, J.R.(1993).*myo*-inositol transport and metabolism in fetal-bovine aortic endothelial cells. *Biochem. J.* **295**, 863–869.

13. Groenen P.M., Peer, P.G., Wevers, R.A., Swinkels, D.W., Franke B, Mariman, E.C., et al. (2003) Maternal *myo*-inositol, glucose, and zinc status is associated with the risk of offspring with spina bifida. *Am. J. Obstet. Gynecol.* **189**, 1713–1719.

14. Greene, D.A., De Jesus, P.V. Jr., and Winegrad, A.I.. (1975) Effects of insulin and dietary *myo*-inositol on impaired peripheral motor nerve conduction velocity in acute streptozotocin diabetes. *J. Clin. Invest.* **55**, 1326–1336.

15. Clements, R.S. Jr., and Darnell, B. (1980). *Myo*-inositol content of common foods: development of a high-*myo*-inositol diet. *Am. J. Clin. Nutr.* **33**, 1954–1967.

16. Spector, R., and Lorenzo, A.V. (1975) *Myo*-inositol transport in the central nervous system. *Am. J. Physiol.* **228**, 1510–1518.

17. Grafton, G., Bunce, C.M., Sheppard, M.C., Brown, G., and Baxter, M.A. (1992) Effect of Mg^{2+} on Na(+)-dependent inositol transport. Role for Mg^{2+} in etiology of diabetic complications. *Diabetes* **41**, 35–39.

18. Eagle, H., Oyama, V.I., Levy, M., and Freeman, A. (1956) *Myo*-inositol as an essential growth factor for normal and malignant human cells in tissue culture. *Science* **123**, 845–847.

19. Eagle, H., Oyama, V.I., Levy, M., and Freeman,A. (1957) *Myo*-inositol as an essential growth factor for normal and malignant human cells in tissue culture. *J. Biol. Chem.* **226**, 191–205.

20. Stephens, L.R., Hawkins, P.T., Morris, A.J., and Downes, P.C. (1988) 1-*myo*-inositol 1,4,5,6-tetrakisphosphate (3-hydroxy)kinase. *Biochem J.* **249**, 283–292.

21. Divecha, N., Banfic, H. and Irvine, R.F. (1991) The polyphosphoinositide cycle exists in the nuclei of Swiss 3T3 cells under the control of a receptor (for IGF-I) in the plasma membrane, and stimulation of the cycle increases nuclear diacylglycerol and apparently induces translocation of protein kinase C to the nucleus. *EMBO. J.* **10**, 3207–3214.

22. Clarke, J. H., Letcher, A. J., D' Santos., C. S., Halstead, J. R., Irvine, R. F. and Divecha, N. (2001) Inositol lipids are regulated during cell cycle progression in the nuclei of murine erythroleukaemia cells. *Biochem. J.* **357**, 905–910.

23. Morris, J. B., Hinchliffe, K. A., Ciruela, A., Letcher, A. J. and Irvine, R. F. (2000) Thrombin stimulation of platelets causes an increase in phosphatidylinositol 5-phosphate revealed by mass assay. *FEBS Lett.* **475**, 57–60.

24. Palmer S., Hughes, K.T., Lee, D. Y, and Wakelam, M.J. (1989) Development of a novel, Ins(1,4,5)P3-specific binding assay. Its use to determine the intracellular concentration of Ins(1,4,5)P3 in unstimulated and vasopressin-stimulated rat hepatocytes. *Cell. Signal.* **1**, 147–156.

25. Mayr, G. W. (1988) A novel metal-dye detection system permits picomolar-range h.p.l.c. analysis of inositol polyphosphates from non-radioactively labelled cell or tissue specimens. *Biochem J.* **254**, 585–591.

26. Maccallum, S. H., Barker, C. J., Hunt, P. A., Wong, N.S., Kirk, C. J., and Michell, R. H. (1989) The use of cells doubly labelled with [14C]inositol and [3H]inositol to search for a hormone-sensitive inositol lipid pool with atypically rapid metabolic turnover. *J Endocrinol.* **122**, 379–389.

27. Menniti, F. S., Oliver, K. G., Nogimori, K., Obie, J. F., Shears, S.B., and Putney, J.W. Jr. (1990) Origins of *myo*-inositol tetrakisphosphates in agonist-stimulated rat pancreatoma cells. Stimulation by bombesin of myo-inositol 1,3,4,5,6-pentakisphosphate breakdown to myo-inositol 3,4,5,6-tetrakisphosphate. *J. Biol. Chem.* **265**, 11167–11176.

28. Staddon, J.M., Barker, C. J., Murphy, A. C., Chanter, N., Lax, A. J., Michell, R. H., et al. (1991) Pasteurella multocida toxin, a potent mitogen, increases inositol 1,4,5-trisphosphate and mobilizes Ca²⁺ in Swiss 3T3 cells. *J. Biol. Chem.* **266**, 4840–4847.

29. Barker, C. J., Wright, J., Hughes, P. J., Kirk, C. J., and Michell, R. H. (2004) Complex changes in cellular inositol phosphate complement accompany transit through the cell cycle. *Biochem J.* **380**, 465–473.

30. York, J.D. (2006) Regulation of nuclear processes by inositol polyphosphates. *Biochim. Biophys. Acta.* **1761**, 552–559.

31. Cleaver, J. E., Thomas, G. H., and Burki, H. J. (1972).Biological damage from intranuclear tritium: DNA strand breaks and their repair. *Science.* **177**, 996–998.

32. Barker, C.J., Wong, N.S., Maccallum, S.M., Hunt, P.A., Michell, R.H., and Kirk, C.J. (1992) The interrelationships of the inositol phosphates formed in vasopressin-stimulated WRK-1 rat mammary tumour cells. *Biochem J.* **286**, 469–474.

33. Howard, C. F. and Anderson, L. (1967) Metabolism of *myo*-inositol in animals. II. Complete catabolism of *myo*-inositol-¹⁴C by rat kidney slices. Arch. Biochem. Biophys. **118**, 332–339

34. Christensen, S. C., Kolbjørn, Jensen, A, Simonsen, L. O. (2003) Aberrant 3H in Ehrlich mouse ascites tumor cell nucleotides after in vivo labeling with myo-[2-3H]- and 1-*myo*-[1-3H]inositol: implications for measuring inositol phosphate signaling. *Anal. Biochem.* **313**, 283–291.

35. Arner, R. J., Prabhu, K. S., Krishnan, V., Johnson, M.C., and Reddy, C. C. (2006) Expression of *myo*-inositol oxygenase in tissues susceptible to diabetic complications. *Biochem. Biophys. Res. Commun.* **339**, 816–20.

36. Hipps, P.P., Holland, W.H., and Sherman, W. R. (1977) Interconversion of myo- and scyllo-inositol with simultaneous formation of neo-inositol by an NADP+ dependent epimerase from bovine brain. *Biochem. Biophys. Res. Commun.* **77**, 340–346.

37. Pak, Y., Huang, L.C., Lilley, K.J., and Larner, J. (1992) In vivo conversion of [³H]*myo*-inositol to [³H]*chiro*-inositol in rat tissues. *J. Biol. Chem.* **267**, 16904–16910.

38. Sun TH, Heimark DB, Nguygen T, Nadler JL, Larner J. (2002) Both *myo*-inositol to chiro-inositol epimerase activities and *chiro*-inositol to *myo*-inositol ratios are decreased in tissues of GK type 2 diabetic rats compared to Wistar controls. *Biochem. Biophys. Res. Commun.* **293**, 1092–1098.

39. Larner J. (2001) D-*chiro*-inositol in insulin action and insulin resistance-old-fashioned biochemistry still at work. *IUBMB Life.* **51**, 139–148.

40. Menniti, F.S., Miller, R.N., Putney, J.W. Jr., and Shears, S.B. (1993) Turnover of inositol polyphosphate pyrophosphates in pancreatoma cells. *J Biol Chem.* **268**, 3850–3856.

41. Stephens, L., Radenberg, T., Thiel, U., Vogel, G., Khoo, K.H., Dell, A. et al. (1993) The detection, purification, structural characterization, and metabolism of diphosphoinositol pentakisphosphate(s) and bisdiphosphoinositol tetrakisphosphate(s). *J. Biol. Chem.* **268**, 4009–4015.

42. Bennett, M., Onnebo, S.M., Azevedo, C., and Saiardi, A. (2006) Inositol pyrophosphates: metabolism and signaling. *Cell. Mol. Life. Sci.* **63**, 552–564.

43. McVeigh, I., and Bracken, E. (1955) The nutrition of Schizosaccharomyces pombe. *Mycologia* **47**, 13–25.

44. Culbertson, M.R., and Henry, S.A..(1975) Inositol-requiring mutants of *Saccharomyces cerevisiae*. *Genetics* **80**, 23–40.

45. Wright, E.M., and Turk, E. (2004) The sodium/glucose cotransport family SLC5. *Pflugers Arch.* **447**, 510–518.

46. Eisenberg, F. Jr., (1967) D-*myo*inositol 1-phosphate as product of cyclization of glucose 6-phosphate and substrate for a specific phosphatase in rat testis. *J. Biol. Chem.* **242**, 1375–1382.

47. Novak, J.E., Turner, R.S., Agranoff, B.W., and Fisher, S.K. (1999) Differentiated human NT2-N neurons possess a high intracellular content of myo-inositol. *J. Neurochem.* **72**, 1431–1440.

48. Wong, Y.H., Kalmbach, S.J., Hartman, B.K., and Sherman, W.R. (1987) Immunohistochemical staining and enzyme activity measurements show *myo*-inositol-1-phosphate synthase to be localized in the vasculature of brain. *J Neurochem.* **48**, 1434–1442.

49. Li, W., Chan, L.S., Khatami, M., and Rockey, J.H. (1986) Non-competitive inhibition of *myo*-inositol transport in cultured bovine retinal capillary pericytes by glucose and reversal by Sorbinil. *Biochim. Biophys. Acta.* **857**, 198–208.

50. Li, W.Y., Zhou, Q., Qin, M., Tao, L., Lou, M., and Hu, T.S. (1991) Reduced absolute rate of myo-inositol biosynthesis of cultured bovine retinal capillary pericytes in high glucose. *Exp. Eye Res.* **52**, 569–73.

51. Del Monte, M.A., Rabbani, R., Diaz, T.C., Lattimer, S.A., Nakamura, J., Brennan, M.C. et al. (1991) Sorbitol, *myo*-inositol, and rod outer segment phagocytosis in cultured hRPE cells exposed to glucose. In vitro model of myo-inositol depletion hypothesis of diabetic complications. *Diabetes* **40**, 1335–1345.

52. Nunez, L.R, and Henry, S.A. (2006) Regulation of 1D-myo-inositol-3-phosphate synthase in yeast. *Subcell. Biochem.* **39**, 135–156.

53. Guan, G., Dai, P., and Shechter, I. (2003) cDNA cloning and gene expression analysis of human myo-inositol 1-phosphate synthase. *Arch Biochem Biophys.* **417**, 251–259.

54. Shamir, A., Shaltiel, G., Greenberg, M.L., Belmaker, R.H., and Agam, G. (2003) The effect of lithium on expression of genes for inositol biosynthetic enzymes in mouse hippocampus; a comparison with the yeast model. *Brain Res. Mol. Brain Res.* **115**, 104–110.

55. Eagle, H., Agranoff, B.W., and Snell, E.E. (1960) The biosynthesis of meso-inositol by cultured mammalian cells, and the parabiotic growth of inositol-dependent and inositol-independent strains. *J. Biol. Chem.* **235**, 1891–1893.

56. Kao, F.T., and Puck, T.T. (1968) Genetics of somatic mammalian cells, VII. Induction and isolation of nutritional mutants in Chinese hamster cells. *Proc. Natl. Acad. Sci. U S A.* **60**, 1275–1281.

57. French, P.J., Bunce, C.M., Stephens, L.R., Lord, J.M., McConnell, F.M., Brown, G. et al. (1991) Changes in the levels of inositol lipids and phosphates during the differentiation of HL60 promyelocytic cells towards neutrophils or monocytes. *Proc Biol. Sci.* **245**, 193–201.

58. Larsson, O., Barker, C.J., Sjöholm, A., Carlqvist, H., Michell, R.H., Bertorello, A. et al. (1997) Inhibition of phosphatases and increased Ca²⁺ channel activity by inositol hexakisphosphate. *Science.* **278**, 471–474.

59. Nogimori K, Menniti FS, Putney JW Jr. (1990) Identification in extracts from AR4-2J cells of inositol 1,4,5-trisphosphate by its susceptibility to inositol 1,4,5-trisphosphate 3-kinase and 5-phosphatase. *Biochem J.* **269**, 195–200.

60. Bird GJ, Oliver KG, Horstman DA, Obie J, Putney JW Jr. (1991) Relationship between the calcium-mobilizing action of inositol 1,4,5-trisphosphate in permeable AR4-2J cells and the estimated levels of inositol 1,4,5-trisphosphate in intact AR4-2J cells. *Biochem J.* **273**, 541–546.

61. Hajra AK, Seguin EB, Agranoff BW. (1968) Rapid labeling of mitochondrial lipids by labeled orthophosphate and adenosine triphosphate. *J. Biol. Chem.* **243**, 1609–1616.

62. Wong, N.S., Barker, C.J., Shears, S.B., Kirk, C.J., and Michell, R.H. (1988) Inositol 1:2(cyclic), 4,5-trisphosphate is not a major product of inositol phospholipid metabolism in vasopressin-stimulated WRK1 cells. *Biochem J.* **252**, 1–5.

63. Lin H, Fridy PC, Ribeiro AA, Choi JH, Barma DK, Vogel G, Falck JR, Shears SB, York JD,

Mayr GW. (2009) Structural analysis and detection of biological inositol pyrophosphates reveal that the family of VIP/diphosphoinositol pentakisphosphate kinases are 1/3-kinases. *J. Biol. Chem.* **284**, 1863–1872.

64. Mulugu, S., Bai, W., Fridy, P.C., Bastidas, R.J., Otto, J.C., Dollins, D.E. et al. (2007) A conserved family of enzymes that phosphorylate inositol hexakisphosphate. *Science* **316**, 106–109.

Chapter 2

HPLC Separation of Inositol Polyphosphates

Christopher J. Barker, Christopher Illies, and Per-Olof Berggren

Abstract

High performance liquid chromatography (HPLC) is an essential analytical tool in the study of the large number of inositol phosphate isomers. This chapter focuses on the separation of inositol polyphosphates from [^{3}H]*myo*-inositol labeled tissues and cells.

We review the different HPLC columns that have been used to separate inositol phosphates and their advantages and disadvantages. We describe important elements of sample preparation for effective separations and give examples of how changing factors, such as pH, can considerably improve the resolving ability of the HPLC chromatogram.

Key words: High performance liquid chromatography, Inositol phosphates, Acid extraction, HPLC columns, HPLC gradients

1. Introduction

High performance liquid chromatography (HPLC) is the method of choice to study inositol polyphosphates: why is it so? The answer can be arrived at, without too much additional text, by looking at the plethora of inositol polyphosphate isomers depicted in Fig. 1. Because of the structure of *myo*-inositol, when its six hydroxyl groups are substituted with, for example, phosphate groups, each position becomes stereochemically unique. Therefore, there is an intrinsic capacity for the construction of a vast catalog of different inositol phosphate isoforms, and this is before you consider pyrophosphorylation. Although not all theoretically possible isomers are found in cells, a high proportion is present (1–5).

Unlike the inositol lipids, isomer-specific mass assays are not generally available for detection of the majority of inositol phosphates. Two exceptions are, inositol 1,4,5-trisphosphate

Christopher J. Barker (ed.), *Inositol Phosphates and Lipids: Methods and Protocols*, Methods in Molecular Biology, vol. 645, DOI 10.1007/978-1-60327-175-2_2, © Humana press, a part of Springer Science+Business Media, LLC 2010

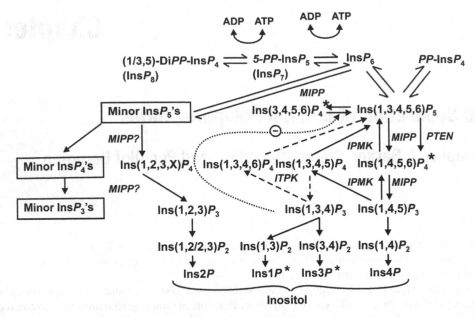

Fig. 1. Multiplicity of inositol phosphate isomers found in cells. The figure demonstrates the complex interrelationships and bewildering array of inositol phosphates. One main issue for HPLC is the fact that the standard anion-exchange chromatography cannot resolve optical isomers. Inositol phosphates with enantiomeric forms are denoted with *.

($Ins(1,4,5)P_3/IP_3$) (6) and the very recently developed assay for inositol hexakisphosphate ($InsP_6$), recorded in this volume (see Chapter 4). Therefore, most workers need to use HPLC to separate this array of isomers to have some confidence that they are indeed measuring one inositol phosphate isoform over another. It is, of course, possible to use the metal-dye detection technique developed by Mayr (7), and presented here in this volume is most recent incarnation (see Chapter 7), without resorting to labeling cells. This technique is commended for those workers whose cellular material is not limiting and who have the resources to establish this assay. However, this technique requires more specialized equipment than available to many workers and is less effective when studying inositol phosphates with less than three phosphate groups. Thus, HPLC separation of [³H]*myo*-inositol-labeled cell extracts is still the main workhorse used to measure the entire spectrum of inositol phosphates present in cells.

1.1. Labeling and Extraction of Cells

A detailed discussion on cell labeling can be found in Chapter 1, and the pitfalls mentioned there should be duly noted before carrying out HPLC on labeled inositol phosphates. In this section, we only refer to the elements of labeling and extraction directly relevant for the HPLC separation of inositol phosphates. Essentially, cells or tissues are prelabeled with a radioisotope, usually [³H]*myo*-inositol. Ideally, and especially for the purposes of trying to relate changes in the amount of a labeled inositol phosphate with changes in its cellular mass, equilibrium labeling is required (see Chapter 1). However, short-term labeling (e.g. 4 h) is useful when assessing inositol phosphate turnover (8–10). Equilibrium labeling needs to be determined empirically (see guidelines in Chapter 1). An important issue with equilibrium labeling is that different inositol phosphates come to equilibrium at different times, with $InsP_6$ and its metabolites labeling most slowly (10). However, from an HPLC analysis perspective, this differential labeling can be used to the investigator's advantage when attempting to measure inositol phosphate isomers derived directly from inositol lipids. This is because you can preferentially label inositol phosphates derived from PLC breakdown of inositol lipids while minimizing the labeling of interfering isoforms (see Note 1). The tendency for investigators to label in conditions of low inositol, for both sensitivity and economic reasons, can confer a similar advantage as long as care is taken in the interpretation (see Chapter 1). For example, $InsP_6$ metabolites, such as $Ins(1,2,3)P_3$, are reduced in concentration under low inositol conditions (11) and thus will interfere less with the immediate products of second messengers derived from G-protein-coupled receptors (e.g. $Ins(1,4,5)P_3$ and $Ins(1,3,4,5,)P_4$). Of course, if you want to measure higher inositol phosphate levels, long-term labeling is essential.

Extraction techniques, which are reviewed in Chapter 1, are important to consider. To recap, inositol phosphates are generally quantitatively extracted using acidic conditions. This has an important bearing on the subsequent chromatography and especially the identification of inositol phosphate isomers. For example, acid extraction will always degrade cyclic inositol polyphosphates and some labile lipid species (see Chapter 1). Acidic degradation of cyclic inositol phosphates will lead to the production of two forms, so for example $Ins(c1:2,4,5)P_3$ will be degraded into $Ins(1,4,5)P_3$ and $Ins(2,4,5)P_3$, similarly for $Ins(c1:2,4)P_2$. Most cell types have relatively little cyclic inositol polyphosphates (12), but in some, they are more abundant (13). Thus, under acidic conditions, an inositol phosphate with a phosphate in the D-2 position could be either a breakdown product of a cyclic inositol phosphate or an $InsP_6$ metabolite. To avoid breakdown of cyclic inositol phosphates, effective neutral extraction strategies have been developed, for example based on a phenol/chloroform saturated phosphate buffer (12).

The amount of extracted cellular material is also important for HPLC analysis. For example, there can be changes in elution position when you move from separating an extract from a 3.5-cm diameter cell culture plate to one that is from a 9-cm plate, especially during isocratic periods of elution. In general, the retention times become shorter the more material that is loaded, and this negatively effects resolution and column lifetime.

The most common form of HPLC column used to separate inositol phosphates is based on anion-exchange, a natural partner with the negatively charged inositol phosphate species (see Note 2). This kind of chromatography, however, will not discriminate between enantiomeric forms (optical isomers) (see Fig. 1). Thus, absolute assignment of structure is not possible with those inositol phosphates that exist as enantiomers. The most important inositol phosphates affected by this issue is the enantiomeric pair $Ins(1,4,5,6)P_4$ and $Ins(3,4,5,6)P_4$ (5, 14). $Ins(3,4,5,6)P_4$ is a second messenger (15), whereas $Ins(1,4,5,6)P_4$s role is unclear. These inositol phosphates are impossible to separate without some kind of chiral-based column or involved chemical degradation techniques (see Note 3). However, enzymatic strategies might now be applied, e.g. see Chapter 3. Historically, a number of different HPLC columns have been used, although there has been a dominance of Whatman columns. Initial studies, when the number of isomeric forms of inositol phosphates started to expand rapidly, used the Whatman Partisil 10 SAX column (16), a column with a 10-μm sized support phase, in its 25 cm form (16). This column is still in limited use today; however, it has largely been superseded by its successor the Whatman Partisphere SAX (a 5-μm column) and the weak anion-exchange variant (Partisphere WAX, also a 5-μm column) (17) (see Note 2). The drop in particle size increased the resolving power (by increasing the number of theoretical plates) and the material was also more regular in structure, hence the suffix, "sphere". The Partisphere columns are available in both 12.5 and 25 cm lengths, but our laboratory favors the 25 cm forms for their greater resolving power. The early advantage of the Partisphere WAX column was its ability to separate metabolically important $InsP_4$ isomers $Ins(1,3,4,5,)P_4$, $Ins(1,3,4,6)P_4$, and $Ins(3,4,5,6)P_4$ (17), a property also possessed by an Altech SAX column (18, 19), probably the other major column used (see Note 4). The referral to these columns in the past tense is not coincidental since both have now been removed from the market. Fortunately, Len Stephens, experimenting with the Partisphere columns, had already demonstrated that increasing the pH from 3.7 to 4.4 produced a reordering of the different isomers of each class and thus could be used to separate the $InsP_4$ isomers of interest (e.g. (20)).

As a consequence of what we have discussed so far, we now need to pay some attention to the elution buffers and at this point it is worthwhile emphasizing that pH changes can dramatically affect the elution positions of inositol phosphates in a particular class. Thus, care must be taken in assigning peak identities when moving from one pH to another and on a practical note, accurate control and measurement of pH are critical to be confident of peak assignments. Figures 2–4 illustrate, for example, the difference in elution position of the $InsP_4$s between pH 3.8 and 4.4. There have been a number of different solvent systems used to elute inositol phosphates from anion exchange columns although phosphate seems to be an important component of most. Originally, concentrations of up to 3.5 M ammonium formate were used with Partisil SAX columns (21), this formate was pH adjusted to 3.7 or 3.8 with orthophosphoric acid. However, if you had the column hooked up to an UV/visible detector, the drastic effect of this high formate concentration on the column was immediately apparent by the large peak of column material that eluted at high-salt concentrations. Ammonium or sodium phosphate gradients have, however, become the dominant elution salts for inositol phosphate and lipid analysis with HPLC, and in our experience, the HPLC column does not suffer unduly with ammonium phosphate (see Note 5). One obvious disadvantage of all these eluants is the fact that online phosphate detection cannot be used with such phosphate-based buffers, and recovery of the inositol phosphate for further analysis is also hampered by the presence of this inorganic phosphate although methods for doing this successfully have been used (e.g. (5, 14) and see Chapter 5). For example, alternatives, which have been used successfully, are eluants based on triethylamine bicarbonate (see Note 6). In Chapter 5, Saiardi's group describes the use of this reagent in desalting inositol phosphates.

We have used the Partisphere SAX (25 cm) column with ammonium phosphate ($[NH_4]_2HPO_4$) adjusted to pH 4.4 with orthophosphoric acid for many of our studies (10, 22). In fact, the Partisphere SAX column offers several advantages over the Partisphere WAX variant, not least its current availability. It is more reproducible, and we have found that it can routinely retain its resolving power up to 200 runs (dependent a little on the individual column). The gradient for the WAX column needed to be reprogrammed every few runs to maintain a viable separation. The disadvantage of strong anion exchange columns is, however, the much higher salt concentrations that are needed to elute the very highly charged inositol phosphates from $Ins(1,3,4,5,6)P_5$ onwards and including the pyrophosphates. This means both that specialized high-salt scintillants are required to detect the radioactivity efficiently in the high-salt environment and that the recovery of the inositol phosphate after chromatography is harder to achieve.

1.3. Detection of Inositol Phosphates and Peak Assignments

In our laboratory, because we are routinely studying less abundant inositol polyphosphates in limiting primary material, we favor (as does the author of the chapter on inositol lipids) using a fraction collector to collect the eluant after elution from the HPLC column. Other workers (e.g. York group, see Chapter 3) use online (or perhaps more properly "inline") detection. The principal is that the column eluant is mixed with scintillant on-the-fly and the resultant mixture passes through a detector cell coupled to photomultiplier tubes. It is worth considering some advantages and disadvantages of the two techniques. Online detection is considerably less labor intensive and can lead to much faster throughput using automated systems that run overnight, if used in combination with an autoinjector. It is strongly recommended for those looking at measuring enzyme activities or other situations where amount of radioactive material is not limiting and there is a sufficient budget. Its disadvantages are that you need more radioactivity in an experiment to reliably detect peaks as the residence time of the sample in the detector is quite short given the commonly used flow rate of 1 mL/min. When collecting fractions the peaks can be counted as long as required, which we have found necessary when study low abundance inositol phosphates in primary or limited cell material. The other issue related to this is that the resolving power is limited by the method of mixing sample and scintillant on-the-fly. However, this loss of resolution can be compensated for by improving the chromatography (see Note 7), or changing the size of the flow cell. Our main reservation is that the amount of radioactivity going into the front end of an experiment on cells is both very expensive and perhaps exposes the cells to higher levels of radioactivity than might be desired (see Chapter 1). We have also found that we use more scintillant per run than when collecting fractions. The advantages of fraction collecting is that very small fractions can be collected when necessary by using a programmable fraction collector and a "windowing" system, as much of an HPLC profile is background. We have found with the Partisphere SAX column (25 cm form) that this column has a great resolving power and collecting smaller fractions in regions of interest exposes this property of the column (for example see Figs. 4 and 5). In terms of scintillants, certainly in European countries, anything other than biodegradable scintillants is frowned upon from an environmental perspective. However, to have a scintillant which both efficiently takes up high salt and particularly phosphate proved to be something of a challenge. We use Packard (now Perkin-Elmer) Ultima Flo AP, which was specifically devised for the separation of inositol phosphates on ammonium phosphate gradients and road-tested in the Michell laboratory. Once a basic HPLC separation has been achieved, using the examples provided in this chapter, some optimization may be required. This is because often the investigator is interested in just one class of inositol phosphates. To achieve the very best separations, long isocratic periods may be needed, e.g. the separation of inositol monophosphates (17). We have studied inositol pyrophosphates recently where only a separation of InsP_5s and more phosphorylated species is required. In this case, a shorter gradient was devised, which meant even with a fraction collector, quite a few samples can be processed in one day (23); a similar approach is used by the Saiardi laboratory in Chapter 5.

In terms of identification of the inositol phosphates present, the current workers in this field are fortunate that they not only have most of the possible forms been identified but also the enzymes responsible for their formation and metabolism characterized, which means standards can be manufactured (see Chapter 3).

In mammalian cells, exhaustive studies looking at structures initially only revealed the expected players (5). However, despite the thoroughness of this approach, some inositol phosphates remained "hidden" because of the conditions in which the cells were grown. For example, $Ins(1,2,3)P_3$ was missed because the original studies to determine structures were carried out in low inositol media (11). So, on a given HPLC column with a known eluant/pH combination, the identification of the peaks can be fairly confidently predicted by their relative retention times without resorting to detailed structural analysis. As stated before, this will not resolve enantiomers (see Fig. 1), which affects prominently one $InsP_4$, the $InsP_5$s, and most recently probably $InsP_7$ and $InsP_8$. Detailed discussion of these latter molecules can be found in Chapters 3, 5, 6, and 7. In Subheading 3 of this chapter, some guidelines are given for peak assignment, and the potential pitfalls to be aware of are discussed. Our approach has been to use either internal standards derived from labeled cells whose inositol phosphate complement has been structurally verified (11), or the approach of splitting a sample into two and adding a commercially available inositol phosphate to half of the sample (see Fig. 5). However, it must be stressed that coelution on an HPLC column with a known inositol phosphate does not definitively prove that the molecule has that structure (see Note 8).

In closing this section, one salutatory tale is worth bearing in mind, which sheds light on both the assignment of peak identities and how experimental protocols may directly impede the interpretation of HPLC chromatograms from labeled cells. TSH stimulation of a rat thyroid cell line was known to increase inositol polyphosphates, which accumulated as a result of using the lithium block. The inhibition of inositol phosphate phosphatases by lithium is a standard assay for PLC activity in which degradation products of the second messenger, $Ins(1,4,5)P_3$ are trapped (24–26). In a more detailed investigation, these data were seemingly supported by HPLC (27), in that peaks thought to correspond to the downstream metabolites of the $Ins(1,4,5)P_3$ second messenger (e.g. $Ins(1,3,4)P_3$) also accumulated on stimulation in the presence of lithium. Taking these results at face value, it could be inferred that TSH was working via PLC degradation of $PtdIns(4,5)P_2$. Unexpectedly, a detailed investigation revealed that at no time did $Ins(1,4,5)P_3$ rise and this was backed up by a lack of $PtdIns(4,5)P_2$ breakdown (27). It was concluded that the inositol phosphates were derived from higher inositol polyphosphate breakdown and not via classic GPCR/PLC $Ins(1,4,5)P_3$ generation. Hindsight also suggests that the $Ins(1,3,4)P_3$ peak that accumulated was in fact probably a different isomer. However, there is an addendum to this story, which highlights another important aspect of HPLC interpretation. In a more recent investigation, again using HPLC, it was reported that in primary human thyroid cells, TSH acts via its receptor to generate $Ins(1,4,5)P_3$ (28), an observation in direct conflict with the study above. The authors of this newer work ascribe the differences to be due to cell and species differences. However, the lesson we wish to draw here relates to the different labeling conditions in the two studies. In the first study, long-term equilibrium labeling was used ensuring the labeling of both the rapidly turning over lipid derived pools and the more slowly turning pools of higher inositol phosphates. In contrast, in the recent study on human thyroid cells, relatively short-term labeling protocols were used in which the higher inositol phosphates would not be well labeled. So, the first study may have preferentially reported higher inositol phosphate metabolism, whereas the latter PLC-mediated events (see also discussion about equilibrium labeling in Chapter 1). Together these two studies highlight the fact that, although HPLC is a powerful technique, it must be used in conjunction with both careful experimental design and the appreciation that reliance on elution position alone for inositol phosphate assignment may be misleading.

2. Materials

2.1. Cell Labeling

1. Adherent or suspension cells.
2. Cell culture medium, inositol free (see Note 9).
3. Dialysis tubing (1,000 MW cut-off) (SpectroPor) (see Note 10).
4. Fetal calf serum.
5. HEPES-buffered Krebs buffer: 109 mM NaCl, 4.6 mM KCl, 1 mM $MgSO_4$, 0.15 mM Na_2HPO_4, 0.4 mM KH_2PO_4, 35 mM $NaHCO_3$, 2 mM CaCl, 20 mM bovine serum albumin 0.5 mg/ml, and Hepes adjusted to pH 7.4.

2.2. Cell Extraction and Preparation

1. Microfuge tubes.
2. Microfuge (e.g. Eppendorf).
3. 15-ml polypropylene tubes (e.g. Starstedt).
4. Ice-cold TCA solution: 5% trichloroacetic acid containing 250 µg/ml of phytic acid ($InsP_6$) (Sigma).
5. Water-saturated diethyl ether.
6. 0.1 M EDTA pH 7.4 (Sigma).
7. Triethylamine.
8. pH indicator paper.

2.3. Elution Protocols for Separating Inositol Polyphosphates

1. HPLC equipment: A two-pump system with high pressure mixing. We use a Gilson with 321 pumps and a UV/VIS-156 detector. For routine analysis, we do not have the UV detector inline, as it slightly reduces the resolution of closely eluting peaks. The HPLC is fitted with a solvent switching valve

that allows the change of the main salt buffer during a run (see Note 11). The equipment, including fraction collector, is controlled by Gilson software.

2. HPLC column: Whatman Partisphere SAX, column (25 cm × 4.6 mm) (see Note 12).

3. Guard cartridge: Whatman Partisphere SAX guard cartridge (see Note 13).

4. Syringe (usually 1-mL disposable plastic syringe with a non-disposable needle having a Luer fitting).

5. Deionized water: filtered (0.2 μm filter) and degassed (Pump A) (see Note 14).

6. The second pump (Pump B) is used to introduce the salt gradient.

7. Ammonium phosphate buffer 1, pH 3.8: 0.1 M $[NH_4]_2HPO_4$, adjusted to pH 3.8 with orthophosphoric acid filtered and degassed as above.

8. Ammonium phosphate buffer 2, pH 3.8: 1.25 M $[NH_4]_2HPO_4$, adjusted to pH 3.8 with orthophosphoric acid filtered and degassed as above.

9. Ammonium phosphate buffer 1, pH 4.4: 0.1 M $[NH_4]_2HPO_4$, adjusted to pH 3.8 with orthophosphoric acid filtered and degassed as above.

10. Ammonium phosphate buffer 2, pH 4.4: 1.25 M $[NH_4]_2HPO_4$, adjusted to pH 3.8 with orthophosphoric acid filtered and degassed as above.

2.4. Detection of Inositol Phosphates and Their Identification

1. Programmable fraction collector: Ideally with a capacity for up to 220, 6 mL vials, we use a Gilson 202.

2. UV/visible detector for HPLC eluant (applicable if using nucleotide standards).

3. 6-ml scintillation vials: we use Pony vials (Perkin-Elmer).

4. Scintillation fluid with a high-salt capacity: Ultima Flo AP (Perkin-Elmer).

5. Inositol phosphate standards: e.g. $[^3H]Ins(1,4,5)P_3$ (GE Healthcare).

6. Nucleotides: e.g. ATP, ADP, AMP, and GTP.

3. Methods

3.1. Cell Labeling

Cells are labeled in medium of choice. There are four variables that need to be considered: (a) the medium inositol concentration – the higher the inositol, the lesser the amount of label incorporated,

(b) related to this, the inositol content of the serum needs to be considered as this also affects labeling, (c) the amount of radioactivity the cells can tolerate, and (d) which inositol phosphates are being measured. To get information on the whole spectrum of inositol phosphates will require more label and/or cells. Using more than 9–10 cm dishes (approx 6 million cells) may cause problems with subsequent HPLC resolution (see Subheading 1) How much label is needed very much depends on the particular inositol phosphate of interest. To measure all inositol phosphates reliably will usually require larger dishes. The methods described below are for the beta cells we commonly use.

1. Cells are plated into labeling medium containing $10\,\mu Ci/ml$ [³H]inositol and allowed to grow for 120–144 h with one to two medium changes (dependent on individual cell line).

2. Cells are then washed 2× with an HEPES-buffered Krebs buffer containing low glucose (basal glucose varies dependent on beta cell line) for up to 2 h.

3. Cells are stimulated with either glucose or other agonists.

4. The above protocols are obviously specific for beta cells.

3.2. Cell Extraction and Preparation

Quantitative extraction of the inositol phosphates from the cells is important. We describe one based on trichloroacetic acid, which we favor because we have found that we get less interference with subsequent HPLC runs. Useful alternative methods can be found in other chapters within this volume that describe inositol phosphate analysis (see Chapters 3–7).

1. Cells are rapidly killed with ice-cold TCA so that the final concentration of TCA is 5%. This is a compromise between extraction ability and the need to reduce the overall acidity. We use 0.5 ml for 3.5 cm dishes and 1 mL for 6 and 9 cm diameter dishes.

2. With adherent cells it is possible to suck off the medium before killing. With cells in suspension direct quenching using an equal volume of 10% TCA to give a final concentration of 5% is more common (see Note 15). Suspension cells could also be rapidly centrifuged and then the media aspirated, as long as this does not interfere with type of measurements being made. There is a possibility that the stress of the centrifugation step could lead to unwanted changes in inositol lipids or phosphates.

3. After killing the sample is left on ice for 20 min.

4. The cell debris is scraped off the plate using a rubber policeman or related cell scraper and the scraping plus acid transferred to a microfuge tube either 1.5 or 2.0 mL dependent on the volume of the extraction.

5. The plate is then washed with 1% TCA (0.5 mL for 3.5 cm dish, 1 mL for 6 or 9 cm dishes).

6. The combined sample is then centrifuged at $6,000 \times g$ in a microfuge (see Note 16).

7. The supernatant is removed to a new tube on ice, usually a 15-ml polypropylene tube (the type used in cell culture).

8. At this point, it is possible to do a bulk lipid extraction of the pellet, which can serve to normalize the inositol phosphate trace (see Note 17).

9. The supernatant is then washed 4× with water saturated diethyl ether. Low speed centrifugation can be used to separate the two phases, but if you use polypropylene tubes, separation of phases happens rapidly without centrifugation (see Note 18).

10. Finally, the ether is removed and 50 µl of a solution of 0.1 M EDTA pH 7.4 is added and a further addition of triethylamine is performed to bring the sample to complete neutrality. This is assessed using pH paper and very small amounts of the sample.

11. The sample is then lyophilized in a vacuum centrifuge to remove the remaining diethyl ether (see Note 19) and then stored at either –20°C or –70°C. The lower temperature seems to preserve samples for longer, but this is measured in years.

3.3. Elution Protocols for Separating Inositol Polyphosphates

The anion exchange chromatography commonly used to separate inositol polyphosphates and deacylated inositol lipids, unlike conventional HPLC, is fairly insensitive to the injection volume because after injection the sample is concentrated in the guard cartridge (or on the top of the main column in the absence of a guard cartridge) and the inositol phosphates do not move down the column until the salt gradient is applied. This means that you can adjust the size of the loading loop according to the size of sample you have, but you must bear in mind that if the sample contains a lot of salts and cellular material large volumes will interfere with the subsequent chromatography and reduce the lifetime of the column. Because of fluid dynamics, it is not recommended that you inject more than half the volume of the loading loop. So the commonly used loop is a 2 ml one in which we inject maximally 1 ml. In order to achieve this volume, some volume reduction may be necessary with attendant problems (see Note 19). However, a less elegant compromise is to do a "double" injection. In essence, the sample is split into two, and two separate injections are given in rapid succession, while only water is pumping through the system. The elution program constructing the salt gradient is then started after the second injection. The only casualty of this modification is the inositol peak although one could also collect fractions from the beginning of the run.

We routinely use two general HPLC gradient profiles, which are identical in construction (see Table 1) but are run with different pHs, that is pH 3.8 and 4.4. When one is interested in trisphosphates onwards, we would recommend pH 4.4 (see Note 20). pH 3.8 is better for the earlier eluting inositol phosphates.

Table 1
Table showing the construction of the HPLC salt gradient used with both pH 3.8 and 4.4 diammonium hydrogen phosphate buffers

Time (min)	% B Eluant 1. 0.1 M $[NH_4]_2HPO_4$
0	0
2	0
15	10
32	13
32.1	100
65	100
65.1 Switch solvent	Eluant 2 1.25 M $[NH_4]_2HPO_4$
95	24
95.1	56
110	56
110.1	64
120	64
120.1	80
165	100
165.1 Switch solvent	Eluant 3. Deionized water (Wash Step)
195	Stop pumps or continue if further samples are being run

The gradients are constructed with isocratic periods followed by steep steps to the next period. Shallow gradients or isocratic periods are essential for the separation of the different isomers. Interestingly, having a long isocratic period early in the run (e.g. in the monophosphate region) will influence the resolution of later classes of inositol phosphates, even up to $InsP_4$s.

The protocols presented here are of course not the only possible methods, even on this column and with ammonium phosphate as the main eluant. Other chapters on inositol phosphate analysis (e.g. Chapters 3–7) give related protocols that could be used. In particular, this general protocol recorded here is not optimized for the separation of $InsP_7$ isoforms and $InsP_8$.

1. We run the Partisphere SAX column with a flow rate of 1 mL/min. Some important preparatory steps are indicated in the Subheading 2.3 (see Note 12).

2. Before any run the column is washed for at least 30 min with deionized water.

3. If the column is brand new, as well as an initial thorough washing with deionized water (see Note 12), we also run at least two dummy gradients before injecting a sample. The first couple of runs are usually not typical of subsequent runs. Furthermore, if you are changing eluant, even if it is only the same solvent at a different pH, a blank run without sample should be undertaken, again for reasons of reproducibility.

4. The sample is defrosted if stored frozen and internal standards added at this point (if required).

5. The sample is spun in a microfuge at maximum speed for 3 min (see Note 21).

6. The sample is then taken up into a polypropylene medical syringe with a luer type needle. Luer fittings are required for most injection loops based on the rheodyne valve. Glass syringes need to be treated to prevent inositol phosphate binding (e.g. Sigmacote, from Sigma).

7. The sample is then loaded into the injection loop (this may be an automatic process if you are using an autoinjector).

8. The loading loop is rotated to inject the column (this could be an automatic process, see above).

9. The gradient is started and so is the fraction collector if this method is being used. (see Chapter 3 for use of online detectors).

10. We use essentially the same HPLC profile for the two different pHs, pH 3.8 or pH 4.4. However, historically we have used a shorter gradient for pH 4.4 (see Note 22).

11. For historical reasons, we use two different concentrations of the same solvent to cover the whole range of inositol phosphates. This is because our original HPLC equipment could not pump reliably at low enough flow rates to accurately control the gradient at low percentages of the salt elution buffer. This led to a delayed elution in the monophosphate region with associated poor resolution of peaks. With more modern HPLC equipment, this may not be necessary and should be determined empirically for your HPLC equipment.

12. The gradient profile described in Table 1 below is then run, only % of Pump B are shown, obviously, Pump A being reciprocal to this.

13. With time salt can build up at various parts of the equipment. Some pumps are fitted with back-flushing and in this case water should be used to wash the pump heads after each day's HPLC. Any salt build up should be washed away using deionized water.

14. Salt can also build up at the fraction collector head. This will lead to larger than optimal drops, which can affect resolution and if the collector is being used on drop mode it will

significant distort the HPLC profile. We always check the drop size before each run.

15. A common problem is a buildup of air in the pump heads. The cause of this can be leaks in the system, improper degassing, or air is injected during sample loading. An indication of air is an oscillating pump pressure. The prevention of such problems is self-evident; however, removing the air afterwards can be more problematic. The air most commonly builds up in the check values and after the disconnection of the plumping, a gentle banging of the valves and/or pump heads with a metal spanner can help to remove the air.

3.4. Detection of Inositol Phosphates and Their Identification

1. Fractions are collected and scintillant added. The amount needed is dependent on collection volume and salt concentration (see Notes 23 and 24).

2. Peak identification can be by comparison of elution position in a given pH buffer. Figures 2–4 give indication of commonly found inositol phosphates and their positions in different pHs.

3. If an internal standard has been used, this will be ideally of a different isotope from the one used to label the cells or tissue.

4. We do not recommend alternating between a standard run and then a sample run to identify peaks, but if this practice is adopted we recommend combining the standard with a blank or dummy extraction of unlabeled cells of the same amount and kind as used for the sample. This at least controls for the effect of the cell extract on the chromatography.

5. As has been discussed in Subheading 1, one of the main issues following chromatography is to discriminate between those inositol phosphates derived immediately from breakdown of $PtdIns(4,5)P_2$ by PLC, notably, $Ins(1,4,5)P_3$, $Ins(1,3,4,5,)P_4$, $Ins(1,4)P_2$, $Ins(1,3)P_2$, and $Ins(3,4)P_2$, from products of higher inositol phosphate metabolism, and particularly $InsP_6$ metabolites. Obviously, a key issue here is the second messenger, $Ins(1,4,5)P_3$ itself. However, a comparison between GPCR stimulated and control will immediately give an indication as to which peaks are likely to be from GPCR-activation of PLC. Another issue is that [^3H]inositol also has contaminants that run at various points in the chromatogram and could be mistakenly identified as inositol phosphates. These are discussed in Chapter 1 (see Note 25).

Fig. 3. (continued) compared to the one used in Fig. 2 (older columns have shorter retention times). (b) Gro$P$Ins is merged with tail of inositol peak, again a sign of column aging. (c) "?" by the $InsP_5$ isoforms reflects uncertainty of actual isoform at this pH. (d) Large fractions have been collected in $InsP_3$ and $InsP_4$ regions, which mean separation is poorer. (e) "$Ins(1,3,4,6)P_4+$" indicates that other 2-phosphate-containing $InsP_4$s could be contaminating this peak. For examples of separations achievable using this pH and smaller fractions see Figs. 4 and 5.

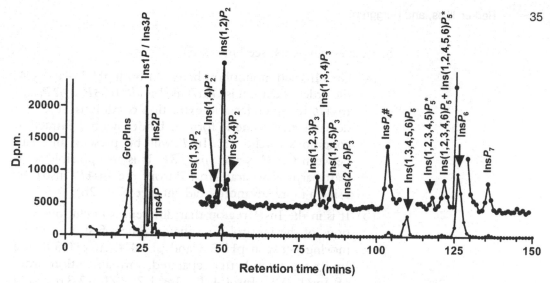

Fig. 2. General HPLC separation of inositol phosphates using pH 3.8 ammonium phosphate buffer. The figure shows a general separation of inositol phosphates using the gradient depicted in Table 1 on a Partisphere SAX column (25 cm × 4.6 mm) fitted with a Partisphere SAX guard cartridge which was eluted with $[NH_4]_2HPO_4$ at pH 3.8. The sample from HIT T15 insulin-secreting cells under basal conditions, which had previously been labeled for 120 h with [³H]inositol, was extracted as described in the methods and injected onto the column. Here it can be seen that GroPIns, InsPs, and the higher inositol phosphates dominate the profile. The inositol peak is off-scale. The smaller peaks are plotted with an expanded scale so that they can be seen more clearly. Positions of known inositol phosphate isomers are depicted. Some inositol phosphates are present as enantiomeric forms and are marked with a/ or a*. The '#' indicates several isomers may be present in the InsP4.

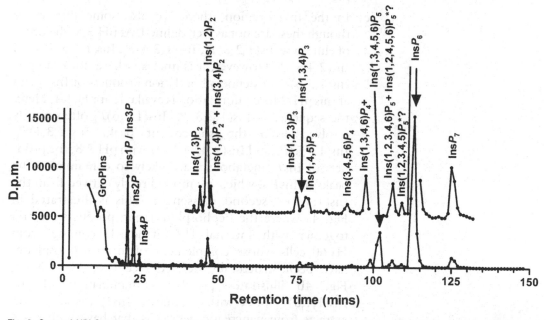

Fig. 3. General HPLC separation of inositol phosphates using pH 4.4 ammonium phosphate buffer. The figure shows a general separation of inositol phosphates using the gradient depicted in Table 1 on a Partisphere SAX column (25 cm × 4.6 mm) fitted with a Partisphere SAX guard cartridge which was eluted with $[NH_4]_2HPO_4$ at pH4.4. The sample from HIT T15 insulin-secreting cells under basal conditions, which had previously been labeled for 120 h with [³H] inositol, was extracted as described in the methods and injected onto the column. Again, the smaller peaks are plotted with an expanded scale so that they can be seen more clearly. Positions of known inositol phosphate isomers are depicted. *Notes* (a) the fact that inositol phosphate peaks run earlier on this chromatogram than the one run at pH 3.8 in Fig. 2 probably does not reflect the difference in pH, but rather the fact that column was of greater age in this figure

6. Peak assignments (see Figs. 2–5)

(a) Gro$P$Ins and monophosphates. At both pH 3.8 and 4.4, the order of elution is Gro$P$Ins, Ins1P/Ins3P, Ins2P, and Ins4P/Ins6P under acid extraction conditions. (the "/" demarks enantiomeric forms). Under neutral extraction conditions, cyclic 1:2 Ins$P$ will be present and run between Gro$P$Ins and Ins1/3P. It is also possible to at least partially separate Ins5P from the Ins4P/Ins6P by using a more advanced gradient (see Note 26).

(b) It is in the InsP_2 region that differences in elution positions of the inositol phosphate isomers first arise when moving between pH 3.8 and pH 4.4. At pH 3.8, the InsP_2s are much better separated, with an elution order of Ins(1,3)P_2, Ins(1,4)P_2, Ins(1,2)P_2/Ins(2,3)P_2, and Ins(3,4)P_3 (which is not so well separated from Ins(1,2)P_2) (see Fig. 2). However, with smaller fractions these peaks can be better separated. Finally, Ins(4,5)P_3 is then eluted. In contrast at pH 4.4, the order is: Ins(1,3)P_3, Ins(1,2)P_3/Ins(2,3)P_3, and then a combined peak of Ins(1,4)P_3 and Ins(3,4)P_3 (see Note 27) and then later Ins(4,5)P_3. Therefore, as far as the separation of InsP_2s is concerned, pH 3.8 is the pH of choice.

(c) In the InsP_3 region, there are also some differences though these are not as well defined. At pH 3.8, the order of elution is Ins(1,2,3)P_3, Ins(1,3,4)P_3, Ins(1,4,5)P_3, and Ins(2,4,5)P_3. However, it is unclear where, for example, Ins(1,5,6)P_3 or dephosphorylation products of InsP_6 run at this pH. More information is available at pH 4.4. Here, the sequence is Ins(1,2,3)P_3, Ins(1,5,6)P_3 often poorly resolved from the subsequent peak, Ins(1,3,4)P_3, Ins(1,4,5)P_3, and Ins(2,4,5)P_3. As with pH 3.8, the products of InsP_6 metabolism are likely to contaminate the peaks of InsP_3s, which are more directly derived from the Ins(1,4,5)P_3 second messenger. This is illustrated in Fig. 4a where the dephosphorylation products are run together with internal [^{14}C]Standards coming from HL60 cells whose complement of inositol phosphates have been structurally defined (see (10, 11)). Note, Fig. 4b illustrates possible contamination of the Ins(1,4,5)P_3 peak with a putative InsP_6 metabolite in extracts from pancreatic beta cells that have high InsP_6 concentrations (29). Clearly, contamination of the Ins(1,4,5)P_3 peak is a serious issue and strategies to combat this are described elsewhere (see Note 28).

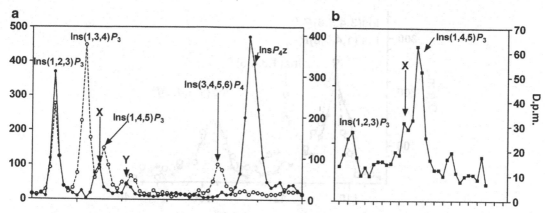

Fig. 4. Contamination of InsP_3 and InsP_4 regions with products of InsP_6 dephosphorylation. Both the following separations were carried out using [NH$_4$]$_2$HPO$_4$ at pH 4.4 and previously published gradients (e.g. (10, 11, 22)). (a) [³H]InsP_6 was dephosphorylated in a homogenate from WRK-1 cells, exactly as described in (11). The resulting inositol phosphates were extracted and combined with [¹⁴C]-standards from HL60 cells. The figure clearly shows that a product of InsP_6 metabolism, "X" runs just in front and poorly resolved from Ins(1,4,5)P_3. Another product "Y" runs after Ins(1,4,5)P_3. (b) Extracts from [³H]inositol-labeled primary beta cells from ob/ob mouse are separated using the gradients referred to above. The chromatogram indicates that a similar peak "X" contaminates the Ins(1,4,5)P_3 peak in normal cell extracts.

(d) In the InsP_4 region, the main inositol phosphates that rise after agonist stimulation are not well separated at pH 3.8. By a careful examination of internal standards (17), the order of elution is Ins(1,3,4,5,)P_4, Ins(3,4,5,6)P_4, and Ins(1,3,4,6)P_4. However, the first two are not separable and the Ins(1,3,4,6)P_4 only poorly separable. On the old Alltech SAX or Whatman Partisphere WAX columns (unfortunately no longer available), all three were well separated (17, 19). However, other InsP_4s, which are downstream from InsP_6 metabolism are separable, both Ins(1,2,3,4)P_4/Ins(1,2,3,6)P_4 and Ins(1,2,5,6)P_4/1ns(2,3,4,5)P_4 elute after the Ins(1,3,4,5,)P_4 peak, with the Ins(1,2,3,4)P_4/Ins(1,2,3,6)P_4 eluting last (17, 20). In order to resolve the common InsP_4s, pH 4.4 can be used. The elution order at this pH is Ins(3,4,5,6)P_4, Ins(1,3,4,5)P_4, and Ins(1,3,4,6)P_4 (see Note 29). The InsP_4 region from an elution run at pH 4.4 is depicted in detail in Fig. 5. This figure also shows the technique of adding a commercial standard, in this case [³H] Ins(1,3,4,5)P_4 to one half of a split sample and then running the samples in two separate runs.

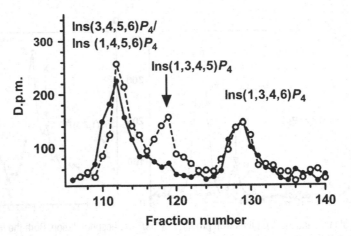

Fig. 5. HPLC separation of InsP_4 region showing the use of an added standard to identify Ins(1,3,4,5)P_4. HIT M2.2.2 insulin-secreting cells were labeled with [^{3}H] inositol and extracted as above. The sample was then split into two and a [^{3}H]Ins(1,3,4,5)P_4 commercial standard added to one half. The samples were separated using a gradient of [NH$_4$]$_2$HPO$_4$ at pH 4.4 as described elsewhere (22). The chromatogram shows the elution position of the major InsP_4 isomers. Note that dephosphorylation products of InsP_6 may lie underneath the Ins(1,3,4,6)P_4 peak (see Fig. 4).

(e) The elution position of the separable InsP_5 isomers at pH 3.8 is as follows: Ins(1,3,4,5,6)P_5, Ins(1,2,3,4,5)P_5/ Ins(1,2,3,5,6)P_5, Ins(1,2,3,4,6)P_5, and Ins(1,2,4,5,6)P_5/ Ins(2,3,4,5,6)P_5. The "/" denotes inseparable enantiomers (e.g. see (30)). Theoretically, this should lead to four possible InsP_5 peaks, but unless using small fractions, it is more difficult to resolve the last two peaks, and normally just three peaks are discerned (see Fig. 2). InsP_6 is a dominant, well-resolved peak although it is possible that it could be contaminated by an InsP_5 pyrophosphate (Di(PP)-InsP_3), this more commonly runs after the InsP_6 peak but before the InsP_7 peak(s). The elution positions of the minor InsP_5s at pH 4.4 have not been formally verified; however, running a similar sample at both pH 3.8 and pH 4.4 seems to reverse the order of the minor InsP_5 isomers (see Figs. 2 and 3).

(f) Note at pH 3.8 the pyrophosphate derivative of Ins(1,3,4,5,6)P_5 runs after the normal InsP_5 isomers, just before the elution of InsP_6. We have not formally verified where this pyrophosphate runs at pH 4.4, but NaF treatment (see below) suggests it runs in the same position as the minor InsP_5s and not close to the InsP_6 peak. For the elution position of inositol pyrophosphates, please see other relevant chapters (e.g. Chapters 3–6). To identify where the pyrophosphates run, preincubation of the cells for 1 h with 5 mM NaF causes an accumulation of pyrophosphate forms (31), which then can be more easily discerned on the HPLC chromatogram. Modified eluants may also be important for pyrophosphate recovery, especially $InsP_8$ (see Note 30).

7. An alternative way of accessing the elution properties of the HPLC column is to use nucleotide standards. AMP runs in the $InsP$ region, ADP in the $InsP_2$ region, ATP in the $InsP_3$ region, and GTP early in the $InsP_4$ region. These can be detected by a UV/Visible spectrophotometer, connected to the column eluant. If nucleotides are included in a run, because the sample goes into a flow cell, this may reduce the resolution of closely eluting inositol phosphate peaks. However, we find the running of nucleotide standards useful to check out the behavior of a column or to improve a gradient profile without having the expense of using labeled inositol phosphates and scintillation fluid.

Overall, over the last 20 years, we have found the Whatman Partisphere SAX columns to be very reproducible batch to batch, with little need to adjust the gradients used. However, recently we have experienced a column which behaved more like the old Partisphere WAX column in the retention times of inositol phosphates (everything came off the column very early) (see Note 31), so it is advisable to always do a test run with a new column before critical samples are inadvertently lost.

4. Notes

1. If you label the total inositol lipids to isotopic equilibrium (usually occurring between 24 and 48 h), there will be a relatively lower labeling of the higher inositol phosphates and their metabolites. This means, for example, in the $InsP_3$ region, contaminating materials from particularly $InsP_6$ metabolism are less likely to interfere with the quantification of the key second messenger, $Ins(1,4,5)P_3$ and its immediate metabolites (e.g. $Ins(1,3,4,5,)P_4$ and $Ins(1,3,4)P_3$). For example, in Fig. 4b, the primary beta cells were labeled for 24 h. Under these conditions, the contaminating peak from $InsP_6$ dephosphorylation does not dominant the $Ins(1,4,5)P_3$ peak.

2. The assignment of SAX (strong anion exchange) is fairly arbitrary; for example, the old Alltech Adsorbosphere SAX column behaved more like the Whatman weak anion exchange (WAX) column in both terms of the ionic strength required for elution of the inositol phosphates and the elution positions of inositol phosphate isomers. So, if you are going to use columns other than those described in Subheading 3 of

this chapter, be aware that you could end up with needing to elute with either stronger or weaker ionic strength and the further hazard of isomers of different inositol phosphate classes changing elution positions.

3. It is possible to resolve these structures by time-consuming degradation methods (e.g. (5, 14, 17, 30)). Although in non-mammalian cells (32] the enantiomer of $Ins(1,4,5)P_3$, $Ins(3,5,6)P_3$ is present, this particular isoform has not been detected in mammalian cells(5).

4. Both the Partisphere WAX and Alltech SAX columns did not, however, give as good a separation of inositol phosphates with less than three phosphate groups.

5. We have no direct experience with the effect of sodium phosphate on column degradation but would anticipate that it would be similar to that of ammonium phosphate.

6. Triethylamine bicarbonate gradients that have been used as the inositol phosphate can be simply de-salted by lyophilization in a freeze drier or a vacuum centrifuge with a good pump and a low temperature trap (e.g. (17)); the principal is also found in Chapter 5 of this book. This technique is not compatible with other commonly used solvent systems as, when mixed with for example a standard elution buffer, carbon dioxide is liberated which will effectively wreck the column.

7. Improvement of the chromatography for online detection will by default lead to longer gradients; however, if the system is automated and running overnight, this should not be too much of an issue. One other important consideration with scintillant/salt mixtures is that if the ratio of scintillant to salt is not high enough irreversible precipitation of inositol pyrophosphates, particularly $InsP_8$ $((PP)_2\text{-}InsP_4)$ can occur (S.B. Shears, personal communication) leading to an underestimate of true pyrophosphate concentrations.

8. An alternative to including standards within a single run is to alternate sample runs with runs of known standards. Our experience is that even with very reproducible chromatography, peaks that elute close to each other cannot be reliably identified using this approach, even if the standards are combined in a blank extraction medium.

9. Dependent on the particular medium some suppliers provide inositol-free versions off-the-shelf. However, this is not always the case. We order custom-made RPMI 1640 from Invitrogen.

10. Some cells do not grow in dialyzed serum, although most will grow in serum dialyzed with a low molecular weight cut-off membrane or tubing. One factor that will be lost is selenium and therefore we advise adding this back before giving up the

use of dialyzed serum in inositol phosphate experiments. Dialyzed FCS can also be obtained commercially.

11. If your HPLC equipment does not have a solvent switching valve, the changeover from one solvent to another can be performed manually. To avoid air getting into the column, the flow should be reduced to zero during the changeover. Once the tubing is placed into the second solvent, it may be possible to purge the pump with the new eluant before restarting the pump. If this is not possible, it is important to increase the isocratic period at this time (for example 5 min) to allow the new stronger eluant to pass through the plumping and reach the column.

12. Columns routinely come in methanol, and this must be thoroughly removed by washing the column with deionized H_2O before the column sees any salt. Mixing the two will destroy the column.

13. We always run our columns with a guard cartridge as this significantly prolongs column life and particularly resolving power by "catching" the general cellular crud that comes from injecting acid cell extracts down the column. We use 3–5 guard cartridges in the lifetime of a single column.

14. We filter all solvents used in HPLC. This in itself seems to be sufficient to degas the aqueous solvents in a two-pump high pressure mixing system. If you have a low pressure mixing system, online degassing is a must and is usually helium based. It is generally the cheaper HPLC systems that have low pressure mixing.

15. The inability to remove the buffer will mean a higher amount of interfering salts reaching the column, with possible changes in elution position of the earlier eluting peaks. The elution position of GroPIns is particularly susceptible to salts and other materials, eluting earlier and closer to the inositol peak.

16. Originally we centrifuged at maximum speed. However, the compact pellet produced at higher g-forces made extracting inositol lipids more difficult. The g-force used is a compromise between having the pellet too loose or too tight and needs to be assessed in each cell system.

17. It is often useful to at least extract total lipids to normalize HPLC traces. To do this, the following simple, modified Bligh–Dyer protocol is recommended. The quantities should be adjusted dependent on the amount of cells being extracted. The amounts below reflect the extraction from confluent 3.5 cm dishes of cells.

(a) Add 0.3 mL of deionized water to the pellet and then 1.124 mL or $CHCl_3$:MeOH: HCl (100:200:1).

(b) Vortex mix and leave on bench for about 15 min.

(c) Add 0.375 mL of $CHCl_3$ and 0.375 mL upper phase mix (1 mM Inositol, 2 M KCl and 0.3 M HCl).

(d) Spin at about $1,000 \times g$ for 5 min at room temperature.

(e) Take off upper phase.

(f) Bottom layer is lipid and this can then be stored under nitrogen or dried down under vacuum. Normally, a small aliquot is first removed and placed into a scintillation vial. The organic solvents are evaporated and the sample combined with scintillant and counted on a liquid scintillation counter. This gives the total inositol lipid in the sample.

(g) A single upper phase wash may be used to remove any remaining contaminating [^{3}H]*myo*-inositol in the lower phase.

18. Diethyl ether is a serious fire risk. It was historically also used as an anesthetic. Therefore, operations involving this solvent should be undertaken under a fume hood, remembering that ether is heavier than air. Furthermore, residues remaining in ether solvent bottles can be explosive and must be handled with caution. The same care should be used when disposing of ether wastes. One should follow local safety regulations in its disposal and handling.

19. It is important to bear in mind that lyophilization of cell extracts can lead to the artificial generation of inositol pyrophosphates ((33) and see Chapter 7). We would thus discourage lyophilizing to dryness and we advice that if this is undertaken the investigator split a sample into two and look at the effect with or without this drying procedure, especially if working on inositol pyrophosphates.

20. pH 4.4 is quite critical for the position of the isomers and the reduction of even 0.1 of a pH unit can start to alter the profile. This is not true running at pH 3.8, where having an accurate pH is more forgiving. Therefore, it is essential to have both a good pH meter and a well-kept probe.

21. In many protocols filters are used before injection. However, early on we discovered that without significant cold carrier there was an almost complete loss of inositol polyphosphates more charged than $InsP_4$ using such filters. Although, use of a cold carrier only becomes an issue if you are going for a rigorous structural analysis of the peaks, on a routine basis we do not use filters. However, without some "filtration step" the expensive column could easily become clogged by particulate matter. This led to the use of a high speed spin in a minifuge (max speed for 3 min) to remove particulate matter that could interfere with the column.

22. We have used a shorter gradient that the one presented here (e.g. (10, 11, 22)). The column has great resolving power so

it is still able to baseline separate inositol phosphate isomers with a shorter gradient, but much smaller fractions need to be collected in regions of interest. Typically 0.1 min fractions.

23. The Ultima-Flo AP was developed, especially for high concentrations of ammonium phosphate and is least efficient with purely aqueous samples. Of course, very high-salt concentrations will need a correspondingly greater proportion of scintillant. The scintillant should easily cope with a 3:1 ratio of scintillant:ammonium phosphate although 4:1 may be advisable in the inositol pyrophosphate region. If the solution is cloudy on addition of scintillant this is an indication that the scintillant has exceeded its capacity. Precipitation can occur which may then lead to less efficient counting of inositol phosphates (see Note 7).

24. One recurrent issue is that of static electricity on plastic vials and the resulting "false" counts this can generate. Many scintillation counters are equipped with antistatic devices that help this. If it does become a problem, often indicated by spurious high counts in just one peak, we suggest the following countermeasures based on our experiences: (a) do not use rubber or latex-based gloves to handle vials, use plastic based ones, (b) recount suspicious vials, (c) wipe vials with methanol (do this if you suspect a spurious value) and recount, and (d) service the scintillation counter.

25. A fact not mentioned in Chapter 1 is that some of the [^{3}H] inositol radiolysis products that can contaminate samples run at the same time irrespective of gradient. This probably indicates they are not interacting with the column in a polar manner but probably with the silica-based support phase. This means that the contaminants will elute under different inositol phosphates on the length of the individual gradients and thus with shorter gradients they may run with more phosphorylated inositol phosphates than the InsP_3s.

26. Stephens et al. devised a long isocratic gradient that could separate all the inositol monophosphates (17). This can now be accomplished by a single 25 cm column rather than two 12.5 cm columns joined together. The 25 cm column was not available at the time of Stephens' original experiments.

27. We have attempted to separate these two InsP_2s using extended isocratic periods of elution. Baseline separation was not achieved, but some separation was possible.

28. If there is concern that there is a contaminating material in the Ins(1,4,5)P_3 peak this can be resolved in a number of different ways. One is to split the sample into two and combine one half with *bona fide* commercially available [^{3}H] Ins(1,4,5)P_3. The samples are then run a gradient with a long isocratic period in the InsP_3 region. A distortion of the peak

shape when comparing the sample with or without spiked $[^3H]Ins(1,4,5)P_3$ will be an indication that there is something wrong. If this is the case, the best route is to check out the $Ins(1,4,5)P_3$ using the commercially available mass assay (e.g. (6, 22)).

29. It is likely that a contaminant from $InsP_6$ dephosphorylation runs in the same position as $Ins(1,3,4,6)P_4$ (e.g. see Fig. 4a).

30. It has been noted that the inclusion of 1 mM EDTA in both aqueous and salt buffers aids the recovery of inositol pyrophosphates (particularly $InsP_8$) from the HPLC column (S.B. Shears, personal communication). Either the free acid or the disodium salt can be used (see also Chapter 5).

31. The defective column, it transpired, was not just a "one-off" issue, the whole lot number; given by the first five digits of the column's serial number was faulty, including 12.5 version of column. For future record, the Lot number to avoid was 8SG05; we hope this was an isolated problem!

Acknowledgments

The authors would like to thank Prof. R.H. Michell and Dr. C.J. Kirk for permission to use data obtained while C.J.B was a research fellow in their laboratory. The authors own work was supported by grants from Karolinska Institutet, Novo Nordisk Foundation, the Swedish Research Council, the Swedish Diabetes Association, EFSD, The Family Erling-Persson Foundation, Berth von Kantzow's Foundation, and EuroDia (LSHM-CT-2006-518153).

References

1. Irvine, R.F., and Schell, M.J. (2001) Back in the water: the return of the inositol phosphates. *Nat. Rev. Mol. Cell. Biol.* 2, 327–338.

2. York, J.D. (2006) Regulation of nuclear processes by inositol polyphosphates. *Biochim. Biophys. Acta.* 1761, 552–559.

3. Shears, S.B. (2007) Understanding the biological significance of diphosphoinositol polyphosphates ('inositol pyrophosphates'). *Biochem. Soc. Symp.* 74, 211–221.

4. Berggren, P.O., and Barker C.J. (2008) A key role for phosphorylated inositol compounds in pancreatic beta-cell stimulus-secretion coupling. *Adv. Enzyme Regul.* 48, 276–294.

5. Wong, N.S., Barker, C.J., Morris, A.J, Craxton, A., Kirk, C.J., and Michell R.H. (1992) The inositol phosphates in WRK1 rat mammary tumour cells. *Biochem. J.* 286, 459–468.

6. Palmer S., Hughes, K.T., Lee, D.Y, and Wakelam, M.J. (1989) Development of a novel, Ins(1,4,5)P3-specific binding assay. Its use to determine the intracellular concentration of Ins(1,4,5)P3 in unstimulated and vasopressin-stimulated rat hepatocytes. *Cell. Signal.* 1, 147–156.

7. Mayr, G.W. (1988) A novel metal-dye detection system permits picomolar-range h.p.l.c. analysis of inositol polyphosphates from non-radioactively labelled cell or tissue specimens. *Biochem. J.* 254, 585–591.

8. Maccallum, S.H., Barker, C.J., Hunt, P.A., Wong, N.S., Kirk, C.J., and Michell, R.H. (1989) The use of cells doubly labelled with [14C]inositol and [3H]inositol to search for a hormone-sensitive inositol lipid pool with atypically rapid metabolic turnover. *J. Endocrinol.* 122, 379–389.

9. Menniti, F.S., Oliver, K.G., Nogimori, K., Obie, J.F., Shears, S.B., and Putney, J.W. Jr. (1990) Origins of *myo*-inositol tetrakisphosphates in agonist-stimulated rat pancreatoma cells. Stimulation by bombesin of myo-inositol 1,3,4,5,6-pentakisphosphate breakdown to myo-inositol 3,4,5,6-tetrakisphosphate. *J. Biol. Chem.* **265**, 11167–11176.

10. Barker, C.J., Wright, J., Hughes, P.J., Kirk, C.J., and Michell, R.H. (2004) Complex changes in cellular inositol phosphate complement accompany transit through the cell cycle. *Biochem. J.* **380**, 465–473.

11. Barker, C.J., French, P.J., Moore, A.J., Nilsson, T., Berggren, P. O., Bunce, C.M., et al. (1995) Inositol 1,2,3-trisphosphate and inositol 1,2- and/or 2,3-bisphosphate are normal constituents of mammalian cells. *Biochem. J.* **306**, 557–564.

12. Wong, N.S., Barker, C.J., Shears, S.B., Kirk, C.J., and Michell, R.H. (1988) Inositol 1:2(cyclic),4,5-trisphosphate is not a major product of inositol phospholipid metabolism in vasopressin-stimulated WRK1 cells. *Biochem. J.* **252**, 1–5.

13. Sekar, M.C., Dixon, J.F., and Hokin, L.E. (1987) The formation of inositol 1,2-cyclic 4,5-trisphosphate and inositol 1,2-cyclic 4-bisphosphate on stimulation of mouse pancreatic minilobules with carbamylcholine. *J. Biol. Chem.* **262**, 340–344.

14. Stephens, L.R., Hawkins, P.T., Morris, A.J., and Downes, P.C. (1988) L-*myo*-inositol 1,4,5,6-tetrakisphosphate (3-hydroxy)kinase. *Biochem. J.* **249**, 283–292.

15. Shears S.B. (1998) The versatility of inositol phosphates as cellular signals. *Biochim. Biophys. Acta.* **1436**, 49–67.

16. Irvine, R.F., Anggård, E.E., Letcher, A.J., and Downes, C.P. (1985) Metabolism of inositol 1,4,5-trisphosphate and inositol 1,3,4-trisphosphate in rat parotid glands. *Biochem. J.* **229**, 505–511.

17. Stephens, L.R., Hawkins, P.T., Barker, C.J., and Downes, C.P. (1988) Synthesis of myo-inositol 1,3,4,5,6-pentakisphosphate from inositol phosphates generated by receptor activation. *Biochem. J.* **253**, 721–733.

18. Balla, T., Guillemette, G., Baukal, A.J., and Catt, K.J. (1987) Metabolism of inositol 1,3,4-trisphosphate to a new tetrakisphosphate isomer in angiotensin-stimulated adrenal glomerulosa cells. *J. Biol. Chem.* **262**, 9952–9955.

19. Hughes, P.J., Hughes, A.R., Putney, J.W. Jr., and Shears, S.B. (1989) The regulation of the phosphorylation of inositol 1,3,4-trisphosphate in cell-free preparations and its relevance to the formation of inositol 1,3,4,6-tetrakisphosphate in agonist-stimulated rat parotid acinar cells. *J. Biol. Chem.* **264**, 19871–19878.

20. Stephens, L.R., Hughes, K.T., and Irvine, R.F. (1991) Pathway of phosphatidylinositol(3,4,5)-trisphosphate synthesis in activated neutrophils. *Nature* **351**, 33–39.

21. Heslop, J.P., Irvine, R.F., Tashjian, A.H. Jr., and Berridge, M.J. (1985) Inositol tetrakis- and pentakisphosphates in GH4 cells. *J. Exp. Biol.* **119**, 395–401.

22. Yu, J., Leibiger, B., Yang, S.N., Caffery, J.J., Shears, S.B., Leibiger, I.B., Barker, C.J., and Berggren, P.O. (2003) Cytosolic multiple inositol polyphosphate phosphatase in the regulation of cytoplasmic free Ca2+ concentration. *J. Biol. Chem.* **278**, 46210–46218.

23. Illies, C., Gromada, J., Fiume, R., Leibiger, B., Yu, J., Juhl, K., et al. (2007) Requirement of inositol pyrophosphates for full exocytotic capacity in pancreatic beta cells. *Science* **318**, 1299–1302.

24. Naccarato WF, Ray RE, Wells WW. (1974) Biosynthesis of myo-inositol in rat mammary gland. Isolation and properties of the enzymes. *Arch. Biochem. Biophys.* **164**, 194–201.

25. Allison, J.H., Blisner, M.E., Holland, W.H., Hipps, P.P., and Sherman, W.R. (1976) Increased brain myo-inositol 1-phosphate in lithium-treated rats. *Biochem. Biophys. Res. Commun.* **71**, 664–670.

26. Berridge, M.J., Downes, C.P., and Hanley, M.R. (1982) Lithium amplifies agonist-dependent phosphatidylinositol responses in brain and salivary glands. *Biochem. J.* **206**, 587–595.

27. Singh, J., Hunt, P., Eggo, M.C., Sheppard, M.C., Kirk, C.J., and Michell, R.H. (1996) Thyroid-stimulating hormone rapidly stimulates inositol polyphosphate formation in FRTL-5 thyrocytes without activating phosphoinositidase C. *Biochem. J.* **316**, 175–182.

28. Van Sande, J., Dequanter, D., Lothaire, P., Massart, C., Dumont, J.E., and Erneux, C. (2006) Thyrotropin stimulates the generation of inositol 1,4,5-trisphosphate in human thyroid cells. *J. Clin. Endocrinol. Metab.* **91**, 1099–1107.

29. Larsson, O., Barker, C.J., Sjöholm, A., Carlqvist, H., Michell, R.H., Bertorello, A., et al. (1997) Inhibition of phosphatases and increased Ca2+ channel activity by inositol hexakisphosphate *Science* **278**, 471–474.

30. Stephens, L.R., Hawkins, P.T., Stanley, A.F., Moore, T., Poyner, D.R., Morris, P.J, et al. (1991) myo-inositol pentakisphosphates.

Structure, biological occurrence and phosphorylation to myo-inositol hexakisphosphate. *Biochem. J.* **275**, 485–499.

31. Menniti, F.S., Miller, R.N., Putney, J.W. Jr., and Shears, S.B. (1993) Turnover of inositol polyphosphate pyrophosphates in pancreatoma cells. *J. Biol. Chem.* **268**, 3850–3856.

32. Stephens, L.R., Berrie, C.P., Irvine, R.F. (1990) Agonist-stimulated inositol phosphate metabolism in avian erythrocytes. *Biochem. J.* **269**, 65–72.

33. Stephens, L., Radenberg, T., Thiel, U., Vogel, G., Khoo, K.H., Dell, A. et al. (1993) The detection, purification, structural characterization, and metabolism of diphosphoinositol pentakisphosphate(s) and bisdiphosphoinositol tetrakisphosphate(s). *J. Biol. Chem.* **268**, 4009–4015.

Chapter 3

Molecular Manipulation and Analysis of Inositol Phosphate and Pyrophosphate Levels in Mammalian Cells

James C. Otto and John D. York

Abstract

Lipid-derived inositol phosphates (Ins$P$s) comprise a family of second messengers that arise through the action of six classes of Ins$P$ kinases, generally referred to as IPKs. Genetic studies have indicated that Ins$P$s play critical roles in embryonic development, but the mechanisms of action for Ins$P$s in mammalian cellular function are largely unknown. This chapter outlines a method for manipulating cellular Ins$P$ profiles through the coexpression of a constitutively active G protein and various IPKs. It provides a mechanism by which the metabolism of a variety of Ins$P$s can be upregulated, enabling the evaluation of the effects of these Ins$P$s on cellular functions.

Key words: Inositol phosphate, G protein, Phospholipase C, HPLC, Second messenger, Signal transduction

1. Introduction

The activation of phospholipase C and subsequent release of inositol 1,4,5-trisphosphate (InsP_3) from phosphatidylinositol 4,5-bisphosphate is a common signaling event in mammalian cells (1, 2). While the role of InsP_3 in calcium signaling has been well established, the lipid-derived InsP_3 is also converted into a variety of more highly phosphorylated species by the action of six classes of inositol phosphate kinase (IPK) (3–6). Higher inositol phosphates (Ins$P$s), which include the inositol diphosphates or pyrophosphates (PP-Ins$P$s), have been shown to play important roles in a variety of cell functions, including regulation of mRNA export, transcription, telomere length, DNA repair/recombination, chromatin remodeling, endocytosis, vacuole biogenesis, and phosphate sensing (4, 5, 7–11). Insight into many of these functions

Christopher J. Barker (ed.), *Inositol Phosphates and Lipids: Methods and Protocols,* Methods in Molecular Biology, vol. 645, DOI 10.1007/978-1-60327-175-2_3, © Humana press, a part of Springer Science+Business Media, LLC 2010

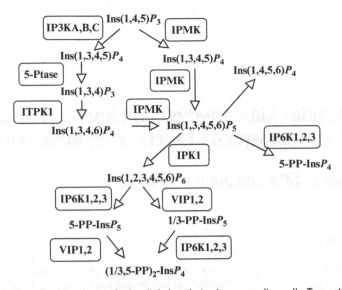

Fig. 1. Biosynthetic pathways for inositol phosphates in mammalian cells. Two pathways exist for the production of InsP_5 in mammalian cells. In the first, IPMK can directly generate InsP_5 from Ins(1,4,5)P_3 by phosphorylating the 3- and 6-position on the inositol ring, moving through an Ins(1,3,4,5)P_4 intermediate. In the second pathway, Ins(1,4,5)P_3 is phosphorylated at the 3-position by an IP3K to generate Ins(1,3,4,5)P_4, which is then dephosphorylated by a 5-phosphatase to generate Ins(1,3,4)P_3. Ins(1,3,4)P_3 is then phosphorylated at the 6-position by ITPK1 to form Ins(1,3,4,6)P_4, which is then phosphorylated at the 5-position by IPMK to generate InsP_5. IPK1 phosphorylates the 2-position of InsP_5 to generate InsP_6. The IP6K family of kinases can phosphorylate the 5-phosphate of InsP_5 and InsP_6, generating inositol pyrophosphates. The VIP kinases also generate inositol pyrophosphates at the 1- or 3-phosphate. The IP6K and VIP kinases can both modify the InsP_7 produced by the other enzyme to generate an InsP_8 with pyrophosphate groups at the 1/3- and 5-phosphates.

has come from studies in the budding yeast model organism. In metazoans, recent studies indicate that production of higher Ins$P$s play critical roles in development, as deletion of the genes for the IPKs that make InsP_5 and InsP_6 results in embryonic lethality (12, 13).

The IPKs and several Ins$P$ phosphatases combine to produce a branched metabolic pathway currently known to extend from InsP_3 to InsP_8 (Fig. 1). The six classes of IPKs in metazoans include the Ins(1,4,5)P_3 3-kinase family (IP3K) (14–16), the Ins(1,3,4)P_3 5/6 kinase (ITPK1) (17, 18), the inositol phosphate multikinase (IPMK/IPK2) essential for InsP_5 production (19–22), the IP$_5$ 2-kinase essential for InsP_6 production (IPK1) (23, 24), the IP6K family that generates inositol pyrophosphates at the 5-phosphate (25–29), and the VIP kinase family that generates pyrophosphates at the 1- and/or 3-phosphate (29–32).

In this chapter, we describe a method to manipulate Ins$P$ metabolism in mammalian cells, leading to the specific production of a variety of Ins$P$s ranging from InsP_3 to InsP_8. This manipulation requires the concerted activation of a G protein along with any one or more of the IPKs. The heterotrimeric G protein subunit $G\alpha_q$ is constitutively activated a Gln209Leu mutation ($G\alpha_q QL$) (33). When expressed in cells, $G\alpha_q QL$ is a potentactivator of phospholipase C and causes a sustained release of InsP_3 (34). By coexpressing $G\alpha_q QL$ with IPKs, the cellular Ins$P$ profile can be dramatically altered. We expect that this ability to remodel the Ins$P$ composition of cells will provide an opportunity to identify new cellular functions for Ins$P$s.

2. Materials

2.1. Isolation and Analysis of Inositol Phosphates from ^{3}H-Inositol Labeled of Cells

1. HEK293T, Rat1, Hela, and Cos7 cells are from Duke Cell Culture Facility (Durham, NC). 12-well culture dishes are from Corning Inc. (Corning, NY).

2. Dulbecco's Modified Eagle Medium (DMEM) (Sigma-Aldrich, St. Louis, MO) is supplemented with 10% fetal bovine serum (Hyclone, Logan, UT) and 1× penicillin/streptomycin (Sigma-Aldrich, St. Louis, MO).

3. Inositol-free DMEM (Specialty Media, Phillipsburg, NJ) is supplemented with 10% dialyzed fetal bovine serum (Invitrogen, Carlsbad, CA), 1× antibiotics, and 37.5 µCi/ml myo-^{3}H-inositol (Perkin Elmer, Waltham, MA).

4. Fugene 6 Transfection Reagent is from Roche Diagnostics, Indianapolis, IN.

5. Phosphate buffered saline (PBS) is from Sigma-Aldrich (St. Louis, MO). HCl is from J.T. Baker (Phillipsburg, NJ).

6. Samples are clarified using a microcentrifuge (Model 5417c; Eppendorf, Westbury, NY).

7. Samples are filtered using a 1.0-ml Luer-Lok syringe (BD Biosciences, San Jose, CA) and 4-mm Acrodisc syringe filters with 0.45-µm nylon membranes (Pall Life Sciences, East Hills, NY).

8. 1.7 M ammonium phosphate (Fisher Scientific, Fairlawn, NJ) is prepared by addition of 784 g of ammonium phosphate to 3 L water. The pH of the solution is adjusted to 3.5 by addition of approximately 7 ml phosphoric acid (EMD Biosciences, Gibbstown, NJ), and the total volume is brought to 4 L. 10 mM ammonium phosphate is prepared by addition of 23.5 ml 1.7 M ammonium phosphate to 4 L of water. The pH of the solution is adjusted to 3.5 by addition of approximately 50 µl phosphoric acid. Ammonium phosphate solutions are sterilized by filtration through 0.22-µm cellulose acetate bottle top filters (Corning Inc., Corning, NY) into autoclaved buffer reservoirs.

9. The components of the HPLC system consist of a System Gold 128 Solvent Module (Beckman Coulter, Fullerton, CA), a 4.6×125 mm Partisphere SAX strong anion column (Whatman, Maidstone, England) and a β–Ram inline radiation detector (IN/US Systems, Tampa, FL). In-Flow BD scintillation fluid (IN/US Systems, Tampa, FL) is used with the radiation detector at a ratio of 2 ml scintillant/ml HPLC eluate.

2.2. Identification of Inositol Phosphates Using Enzymatic Analysis

1. Cell culture reagents are identical to those described in Subheading 2.1. Tris base, Hepes, KCl, $MgCl_2$, KOH, and ATP are from Sigma-Aldrich (St. Louis, MO).

2. A stock solution of 1 M Tris base is prepared in water and adjusted to pH 8.0 using HCl; 50 mM Tris–HCl (pH 8.0) is prepared by diluting this solution in water.

3. A stock solution of 1 M Hepes is prepared in water and adjusted to pH 7.5 by addition of 10 M KOH; 1 M solutions of KCl and $MgCl_2$ are prepared in water. Enzyme reaction buffer is prepared as a 5× stock consisting of 250 mM K-Hepes (pH 7.5), 250 mM KCl, and 50 mM $MgCl_2$.

4. A 50-mM stock solution of ATP is prepared in 100 mM Tris–HCl, pH 8.0. The stock solution is aliquoted into microcentrifuge tubes and stored at –80°C.

5. The expression and purification of *A. thaliana* IPK1, *A. thaliana* IPMK, and human $Ins(1,4,5)P_3$ type I 5-phosphatase (5-Ptase) have been described in detail previously (35, 36). The cDNA for human diphosphoinositol phosphate phosphatase (DIPP) was obtained from Stephen Shears (NIEHS, Research Triangle Park, NC) and the enzyme is expressed and purified as described (37).

6. Materials and equipment for HPLC analysis are identical to those described above.

2.3.–2.8. Upregulation of Inositol Phosphate Metabolism by Expression of $G\alpha_q QL$ and IPKs

1. Cell culture reagents are identical to those described above.

2. The cDNA for human $G\alpha_q QL$ (pcDNA-$G\alpha_q QL$) is from UMR cDNA Resource Center (Rolla, MO). pcDNA3.1 is obtained from Invitrogen (Carlsbad, CA). The cDNA for human ITPK1 (pCMV-SPORT6-hITPK1) and human IP6K1 (pCMV-SPORT6-hIP6K1) are from Open Biosystems (Huntsville, AL). The cDNA for rat IPMK (pBabePuro-GFP-rIPMK), human IPK1 (pcDNA3.1-hIPK1), human ITPKA (pcDNA3.1-CFP-hITPKA), and human VIP1 kinase domain (pmCFP-hVIP1-KD) are as described previously (31, 34, 38).

3. Materials and equipment for HPLC analysis are identical to those described above.

3. Methods

Coexpression of the phospholipase C activator $G\alpha_q QL$ with various IPKs in mammalian cells can be utilized to drive the intracellular production of Ins$P$s ranging from InsP_3 to InsP_8. The protocols described here outline the labeling of mammalian cells with ^{3}H-inositol and analysis of the resulting Ins$P$ metabolic profiles by HPLC (Subheading 3.1), the use of enzymatic analysis to determine the identity of Ins$P$ isomers (Subheading 3.2, summarized in Table 1), and $G\alpha_q QL$/IPK combinations that upregulate the production of specific Ins$P$s (Subheading 3.3–3.8, summarized in Table 2).

3.1. Isolation and Analysis of Inositol Phosphates from ^{3}H-Inositol Labeled of Cells

1. Plate cells at a density of 25,000–100,000 cells/ml on 12-well culture dishes and allow to adhere to the dish overnight.

2. Wash cells in inositol-free DMEM, and label for 72 h in inositol-free DMEM containing 10% dialyzed fetal bovine serum, antibiotics, and 37.5 µCi/ml myo-^{3}H-inositol.

3. Transfect cells 24 h prior to harvest. 100 ng of cDNA for $G\alpha_q QL$ and each IPK are used per transfection, and empty pcDNA3.1 is added to bring the total amount of DNA to 500 ng. 1.5 µl Fugene 6 is mixed with 50 µl DMEM, incubated for 5 min at room temperature, and then mixed with the DNA. Fugene/DNA complexes are incubated 15 min at room temperature, and then added directly into the labeling media.

Table 1
Identification of Ins$P$s by enzymatic analysis

Ins$P$	atIPK1	atIPMK	h5-Ptase
Ins$(1,4,5)P_3$	No	Yes	Yes
Ins$(1,3,4,5)P_4$	No	Yes	Yes
Ins$(1,3,4,6)P_4$	Yes	Yes	No
Ins$(1,4,5,6)P_4$	No	Yes	No
		DIPP (product produced)	
5-PP-InsP_4		InsP_5	
5-PP-InsP_5		InsP_6	
$(1/3,5$-$PP)_2$-InsP_4		InsP_6	

The table indicates the expected outcomes for the enzymatic analysis of various Ins$P$s present in cell extracts. In the case of IPK1, IPMK, and 5-Ptase, the susceptibility of a given Ins$P$ to an enzyme is indicated by a "Yes" or "No". For DIPP, the expected product of each PP-Ins$P$ following DIPP treatment is indicated

Table 2
Induction of distinct Ins$P$s by coexpression of Gα_qQL and IPKs

Ins$P$	Enzyme coexpressed with Gα_qQL
Ins(1,3,4,5)P_4	IP3KA
Ins(1,3,4,6)P_4	ITPK1
Ins(1,4,5,6)P_4	IPMK
Ins(1,3,4,5,6)P_5	IPMK
InsP_6	IPMK+IPK1
5-PP-InsP_4	IP6K1
5-PP-InsP_5 (InsP_7)	IPK1+IP6K
(1/3,5-PP)$_2$-InsP_4 (InsP_8)	IPK1+IP6K+VIP1-KD

The table indicates combinations of IPKs that upregulate the production of specific species of Ins$P$ in mammalian when coexpressed with Gα_qQL

4. Cells are washed with PBS and harvested in 0.5 M HCl (See Note 1).

5. Cell extracts are clarified by centrifugation at $16,000 \times g$ and passed though a 0.45 μm filter.

6. Ins$P$s in cell extracts are analyzed by HPLC. Extracts are diluted with four volumes of 10 mM ammonium phosphate (pH 3.5) and are resolved on a Partisphere SAX strong anion column using a 65-min linear gradient of ammonium phosphate (pH 3.5) ranging from 10 mM to 1.7 M, followed by 1.7 M ammonium phosphate for 40 min. This gradient allows separation of inositol phosphates ranging from InsP_2 to InsP_8.

7. Radiolabeled Ins$P$s are quantified using an inline radiation detector.

3.2. Identification of Inositol Phosphates Using Enzymatic Analysis

1. Cells are labeled with ^{3}H-*myo*-inositol and transfected as described above.

2. Cells are harvested by washing in PBS and then were scraped into 50 mM Tris–HCl (pH 8.0) that has been heated to boiling. Extracts are transferred to a microcentrifuge tube.

3. Extracts are then boiled for an additional 5 min.

4. Cell extracts were passed through a 22-gauge needle several times.

5. Debris is removed by centrifugation at $16,000 \times g$ and the supernatant is passed through a 0.45-μm filter.

6. Extracts are subjected to enzymatic analysis by incubation with 1 μg of enzyme in 50 mM Hepes (pH 7.5), 50 mM KCl, 10 mM $MgCl_2$, and 1 mM ATP in 50 μl at 37°C for 20 min.

7. Reactions are quenched by addition of 50 μl of 0.5 M HCl.

8. Samples are prepared for HPLC analysis by addition of 400 μl 10 mM ammonium phosphate (pH 3.5) and are passed through a 0.45-μm filter.

9. Table 1 outlines the identification of $InsP_4$ and $PP\text{-}InsP$ species by enzymatic analysis.

3.3. Upregulation of InsP Metabolism by Expression of $G\alpha_qQL$ and IPMK

1. HEK293T cells (100,000 cells/well) are labeled with ^{3}H-inositol for 48 h and transfected with either empty vector, IPMK, $G\alpha_qQL$, or $G\alpha_qQL$+IPMK.

2. Cells are harvested, and the metabolic profiles of the InsPs in the extracts are assessed by HPLC (Fig. 2).

3. Expression IPMK alone has minimal effects on the InsP metabolic profile of the cell, while expression of $G\alpha_qQL$ leads to increases in the levels of $InsP_3$ and $InsP_4$ in the metabolic profile.

4. When $G\alpha_qQL$ and IPMK are coexpressed together, large increases in the levels of $InsP_3$, $InsP_4$, and $InsP_5$ are observed.

5. Enzymatic analysis identifies the predominant isomers of $InsP_3$ peak as $Ins(1,4,5)P_3$ and the $InsP_4$ peak as $Ins(1,4,5,6)P_4$ (see Note 2).

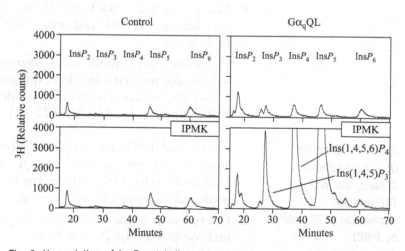

Fig. 2. Upregulation of InsP metabolism through coexpression of $G\alpha_qQL$ and IPMK. HEK293T cells are metabolically labeled with ^{3}H-inositol for 48 h and then transfected with empty pcDNA3.1 (control) or $G\alpha_qQL$ and IPMK, alone and in combination as noted. InsPs are analyzed by HPLC. Isomers of $InsP_3$ and $InsP_4$ produced by coexpression of $G\alpha_qQL$ and IPMK are identified by enzymatic analysis.

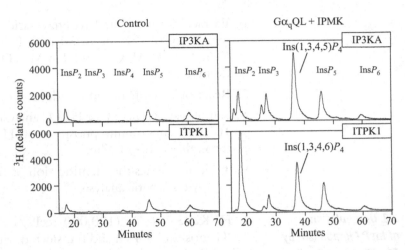

Fig. 3. Coexpression of ITPKA and ITPK1 with $G\alpha_q$QL produce distinct species of $InsP_4$. HEK293T cells are labeled with ^{3}H-inositol and transfected with either ITPK1 or IP3KA alone or in combination with $G\alpha_q$QL (as noted). Inositol phosphates are analyzed by HPLC, and the major species of $InsP_4$ produced by the transfections are identified by enzymatic analysis.

3.4. Generation of Ins(1,3,4,5)P_4 and Ins(1,3,4,6)P_4 by Coexpression of $G\alpha_q$QL and $InsP_3$ Kinases

1. HEK293T cells are labeled with ^{3}H-inositol and transfected with either IP3KA or ITPK1 in the absence or presence of $G\alpha_q$QL.

2. Cells are harvested, and the $InsP$ metabolic profiles are analyzed by HPLC (Fig. 3).

3. Expression of IP3KA and ITPK1 alone has minimal effects on the metabolic profile (Fig. 3).

4. Coexpression of IP3KA and ITPK1 with $G\alpha_q$QL generates large increases in $InsP_4$ and smaller but significant increases in $InsP_5$ levels.

5. Enzymatic analysis of the $InsP_4$ isomers reveals that the major species accumulated following expression of each enzyme was in fact the expected enzymatic product, Ins(1,3,4,5)P_4 for ITPKA and Ins(1,3,4,6)P_4 for ITPK1 (see Note 3).

3.5. Unmasking $G\alpha_q$ QL Induced Fluxes in $InsP_5$ and $InsP_6$ Through the Generation of PP-$InsP_s$ by IP6K1

1. IP6K1 and IPK1 are expressed alone or together in the absence and presence of $G\alpha_q$QL in ^{3}H-inositol-labeled HEK293T cells. Cell extracts are prepared, and metabolic profiles are analyzed by HPLC (Fig. 4).

2. Coexpression of the IP6K1 with $G\alpha_q$QL results in a dramatic increase in 5-PP-$InsP_4$. (see Note 4).

3. Coexpression of IPK1 and $G\alpha_q$QL does not lead to significant changes in the $InsP$ metabolic profile.

4. Coexpression of IP6K1, IPK1, and $G\alpha_q$QL results in a dramatic increase in 5-PP-$InsP_5$ (see Note 5).

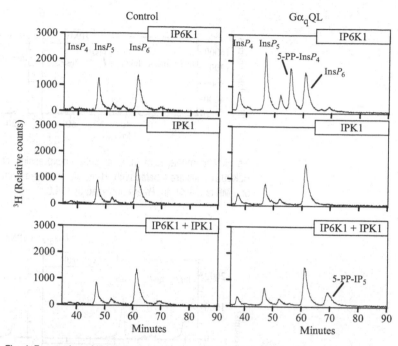

Fig. 4. Expression of IP6K1 unmasks fluxes of InsP_5 and InsP_6 in cells expressing Gα_qQL. HEK293T cells are labeled with ³H-inositol and transfected with IP6K1, IPK1, or IP6K1 and IPK1 both in the absence (*left panels*) and presence (*right panels*) of Gα_qQL. Inositol phosphates are analyzed by HPLC. Coexpression of IP6K with Gα_qQL results in the generation of 5-*PP*-InsP_4, indicating the presence of a flux of InsP_5 in cells expressing Gα_qQL. Coexpression of IPK1 with IP6K and Gα_qQL produces 5-*PP*-InsP_5, indicating that the InsP_5 flux can be shifted to InsP_6 through expression of IPK1.

5. The accumulation of 5-*PP*-InsP_4 in cells expressing IP6K1 with Gα_qQL indicates that a flux of InsP_5 synthesis occurs in cells that are not evident in the Ins$P$ metabolic profile (see Fig. 2). Coexpression of IPK1 and Gα_qQL causes a similar flux in InsP_6 production that can be captured by IP6K1, leading to the accumulation of 5-*PP*-InsP_5 (see Note 6).

3.6. Induction of InsP₆ Synthesis

1. IPMK and IPK1 are coexpressed with Gα_qQL in ³H-inositol-labeled HEK293T cells. Ins$P$ metabolic profiles of cell extracts are analyzed by HPLC (Fig. 5).

2. Coexpression of IPMK, IPK1, and Gα_qQL leads to the accumulation of large levels of InsP_6.

3.7. Generation of (PP)₂-InsP₄ in Mammalian Cells

1. VIP1 kinase domain is either with Gα_qQL and IPK1 or with Gα_qQL, IPK1, and IP6K1. Cells are metabolically labeled with ³H-inositol and transfected with the appropriate constructs.

2. The Ins$P$ metabolic profiles of cell extracts are analyzed by HPLC (Fig. 6).

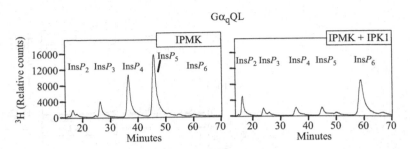

Fig. 5. Accumulation of InsP$_6$ in cells coexpressing of IPMK and IPK1 with Gα_qQL. HEK293T cells are labeled with ^{3}H-inositol and transfected with Gα_qQL and either IPMK or IPMK and IPK1. InsPs are analyzed by HPLC.

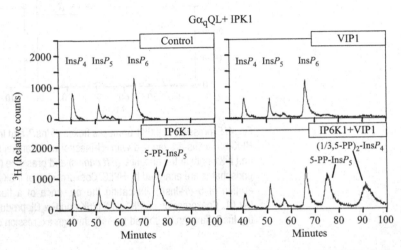

Fig. 6. Generation of InsP$_8$ by coexpression of IP6K1 and VIP1 kinases. HEK293T cells are labeled with ^{3}H-inositol and transfected with Gα_qQL and IPK1 and either control plasmid (*top left*), VIP1 kinase domain (*top right*), IP6K1 (*bottom left*) or IP6K1 and VIP1 kinase domain (*bottom right*). InsPs are analyzed by HPLC. The identity of InsP$_7$ and InsP$_8$ species are noted.

3. No additional InsPs are present in cells expressing VIP1 with IPK1 and Gα_qQL (see Note 7).

4. Coexpression of VIP1 with IPK1, IP6K1, and Gα_qQL leads to the generation of $(1/3,5\text{-}PP)_2\text{-}InsP_4$ (see Note 8).

3.8. Upregulation of InsP Metabolism by Gα_qQL in Additional Cell Lines

1. Rat1 (25,000 cells/well), Hela (100,000 cells/well), and Cos7 (50,000 cells/well) are labeled with ^{3}H-inositol and transfected with Gα_qQL and IPMK.

2. Cell extracts are prepared, and the InsP metabolic profiles are analyzed by HPLC (Fig. 7).

3. Elevated levels of InsP$_3$, InsP$_4$, and InsP$_5$ occur in each of the cell lines (see Note 9).

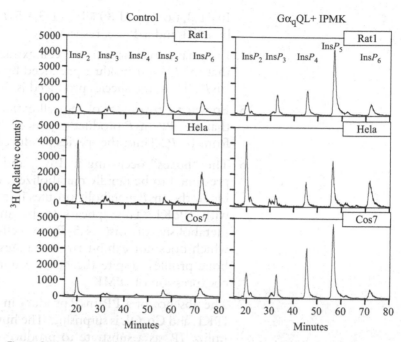

Fig. 7. Coexpression of $G\alpha_q$QL and IPMK in Rat1, Hela, and Cos7 cells. Cell lines are labeled with ^{3}H-inositol and transfected with either empty pcDNA3.1 (*left panels*) or with $G\alpha_q$QL and IPMK (*right panels*). Ins*P*s are analyzed by HPLC.

4. Notes

1. For harvesting Ins*P*s from cells in 0.5 M HCl, it is typically sufficient to soak the cells in the acid for 5 min and then transfer the liquid for further processing. Scraping the cells into the acid and using mechanical lysis does not significantly increase the Ins*P* recovery.

2. Enzymatic analysis indicates that Ins(1,4,5,6)P$_4$ is approximately 49% of the Ins*P*$_4$ present in the cell extract. IPMK converts 82% of the Ins*P*$_4$ to Ins*P*$_5$. 22% of the Ins*P*$_4$ was substrate for IPK1, indicating the presence of 22% Ins(1,3,4,6)P$_4$ in the sample, and 11% of the Ins*P*$_4$ was utilized by 5-Ptase, indicating the presence of 11% Ins(1,3,4,5)P$_4$. The 49% value for Ins(1,4,5,6)P$_4$ is arrived at by subtracting the percentage of the sample susceptible to IPK1 and 5-Ptase from that utilized by IPMK.

3. 86% of the Ins*P*$_4$ produced by coexpression of IP3KA and $G\alpha_q$QL is utilized by 5-Ptase, indicating Ins(1,3,4,5)P$_4$. The Ins*P*$_4$ produced by coexpression of ITPK1 and $G\alpha_q$QL is 60%

Ins$(1,3,4,6)P_4$ and 30% Ins $(1,3,4,5)P_4$, based on utilization by IPK1 and 5-Ptase, respectively.

4. Enzymatic treatment of the cell extract with DIPP indicates that the PP-Ins$P$ product produced by IP6K1 is derived from InsP_5. Thus, the species produced is 5-PP-InsP_4.

5. Enzymatic treatment of the cell extract with DIPP indicates that the PP-Ins$P$ product produced by IP6K1 was derived from InsP_6. Thus, the species produced is 5-PP-InsP_5.

6. The "fluxes" occurring in InsP_5 and InsP_6 production are presumed to be rapidly metabolized, rendering them invisible in the Ins$P$ metabolic profiles in the absence of trapping with IP6K1. This appears to be analogous to the rapid metabolism of Ins$(1,4,5)P_3$ in cells expressing Gα_qQL, which does not exhibit large changes as seen in Ins$P$ metabolic profiles despite the massive upregulation exposed by coexpression of IPMK.

7. The absence of a PP-Ins$P$ product in cells expressing VIP1, IPK1, and Gα_qQL is surprising. The human VIP kinases readily utilize IP$_6$ as a substrate to produce 1/3-PP-InsP_5 in both in vitro assays and when expressed in yeast (31). Thus, although the VIP kinases should be capable of producing InsP_7 from InsP_6 in mammalian cells, at this time the conditions under which it might occur remain unclear.

8. Enzymatic treatment of this extract with DIPP indicates that the PP-Ins$P$ species present are all derived from InsP_6. Therefore, the InsP_8 species produced in the cells is $(1/3,5)$-PP_2-InsP_4.

9. The magnitude of the increases in Ins$P$ metabolism observed appears to be directly related to the transfection efficiency of the cells, with Rat1 cells having the lowest transfection efficiency and demonstrating the smallest increase in Ins$P$ metabolism, and Cos7 cells having the highest transfection efficiency and the most dramatic change. Therefore, it appears that combination of Gα_qQL and IPKs will be useful in upregulating Ins$P$ metabolism in a variety of cell lines.

Acknowledgments

This work was supported by funds from the Howard Hughes Medical Institute and from the National Institute of Health Grants HL-55672 and DK-070272. We thank members of the York laboratory for their contributions to the study of IPKs.

References

1. Berridge, M. J., & Irvine, R. F. (1989) Inositol phosphates and cell signalling. *Nature.* **341**, 197–205.

2. Rhee, S. G. (2001) Regulation of phospho-inositide-specific phospholipase C. *Annu. Rev. Biochem.* **70**, 281–312.

3. Majerus, P. W. (1992) Inositol phosphate biochemistry. *Annu. Rev. Biochem.* **61**, 225–250.

4. Irvine, R. F., Schell, M. J. (2001) Back in the water: the return of the inositol phosphates. *Nat. Rev. Mol. Cell Biol.* **2**, 327–338.

5. Bennett, M., Onnebo, S. M., Azevedo, C., Saiardi, A. (2006) Inositol pyrophosphates: metabolism and signaling. *Cell Mol. Life Sci.* **63**, 552–564.

6. Irvine, R. (2007) Cell signaling. The art of the soluble. *Science.* **316**, 845–846.

7. Shears, S. B. (2004) How versatile are inositol phosphate kinases? *Biochem. J.* **377**, 265–280.

8. York, J. D. (2006) Regulation of nuclear processes by inositol polyphosphates. *Biochimica et biophysica acta.* **1761**, 552–559.

9. Bhandari, R., Chakraborty, A., & Snyder, S. H. (2007) Inositol pyrophosphate pyrotechnics. *Cell Metab.* **5**, 321–323.

10. Onnebo, S. M., Saiardi, A. (2007) Inositol pyrophosphates get the vip1 treatment. *Cell.* **129**, 647–649.

11. Lee, Y. S., Mulugu, S., York, J. D., & O'Shea, E. K. (2007) Regulation of a cyclin-CDK-CDK inhibitor complex by inositol pyrophosphates. *Science.* **316**, 109–112.

12. Frederick, J. P., Mattiske, D., Wofford, J. A., Megosh, L. C., Drake, L. Y., Chiou, S.-T., Hogan, B. L. M., & York, J. D. (2005) An essential role for an inositol polyphosphate multikinase, Ipk2, in mouse embryogenesis and second messenger production. *PNAS.* **102**, 8454–8459.

13. Verbsky, J., Lavine, K., & Majerus, P. W. (2005) Disruption of the mouse inositol 1,3,4,5,6-pentakisphosphate 2-kinase gene, associated lethality, and tissue distribution of 2-kinase expression. *PNAS.* **102**, 8448–8453.

14. Choi, K. Y., Kim, H. K., Lee, S. Y., Moon, K. H., Sim, S. S., Kim, J. W., Chung, H. K., & Rhee, S. G. (1990) Molecular cloning and expression of a complementary DNA for inositol 1,4,5-trisphosphate 3-kinase. *Science.* **248**, 64–66.

15. Irvine, R. F., Letcher, A. J., Heslop, J. P., & Berridge, M. J. (1986) The inositol tris/ tetrakisphosphate pathway – demonstration of Ins(1,4,5)P3 3-kinase activity in animal tissues. *Nature.* **320**, 631–634.

16. Takazawa, K., Lemos, M., Delvaux, A., Lejeune, C., Dumont, J. E., & Erneux, C. (1990) Rat brain inositol 1,4,5-trisphosphate 3-kinase. Ca2(+)-sensitivity, purification and antibody production. *Biochem. J.* **268**, 213–217.

17. Wilson, M. P., & Majerus, P. W. (1996) Isolation of inositol 1,3,4-trisphosphate 5/6-kinase, cDNA cloning and expression of the recombinant enzyme. *J. Biol. Chem.* **271**, 11904–11910.

18. Yang, X., & Shears, S. B. (2000) Multitasking in signal transduction by a promiscuous human Ins(3,4,5,6)P(4) 1-kinase/Ins(1,3,4) P(3) 5/6-kinase. *Biochem. J.* **351**(Pt 3), 551–555.

19. Chang, S. C., Miller, A. L., Feng, Y., Wente, S. R., & Majerus, P. W. (2002) The human homolog of the rat inositol phosphate multikinase is an inositol 1,3,4,6-tetrakisphosphate 5-kinase. *J. Biol. Chem.* **277**, 43836–43843.

20. Nalaskowski, M. M., Deschermeier, C., Fanick, W., & Mayr, G. W. (2002) The human homologue of yeast ArgRIII protein is an inositol phosphate multikinase with predominantly nuclear localization. *Biochem. J.* **366**, 549–556.

21. Odom, A. R., Stahlberg, A., Wente, S. R., & York, J. D. (2000) A role for nuclear Inositol 1,4,5-Trisphosphate Kinase in transcriptional control. *Science.* **287**, 2026–2029.

22. Saiardi, A., Caffrey, J. J., Snyder, S. H., & Shears, S. B. (2000) Inositol polyphosphate multikinase (ArgRIII) determines nuclear mRNA export in *Saccharomyces cerevisiae*. *FEBS Lett.* **468**, 28–32.

23. Verbsky, J. W., Wilson, M. P., Kisseleva, M. V., Majerus, P. W., & Wente, S. R. (2002) The synthesis of inositol hexakisphosphate. Characterization of human inositol 1,3,4,5,6-pentakisphosphate 2-kinase. *J. Biol. Chem.* **277**, 31857–31862.

24. York, J. D., Odom, A. R., Murphy, R., Ives, E. B., & Wente, S. R. (1999) A phospholipase C-dependent inositol polyphosphate kinase pathway required for efficient messenger RNA export. *Science.* **285**, 96–100.

25. Saiardi, A., Caffrey, J. J., Snyder, S. H., & Shears, S. B. (2000) The inositol hexakisphosphate kinase family. Catalytic flexibility and function in yeast vacuole biogenesis. *J. Biol. Chem.* **275**, 24686–24692.

26. Saiardi, A., Erdjument-Bromage, H., Snowman, A. M., Tempst, P., & Snyder, S. H. (1999) Synthesis of diphosphoinositol pentakisphosphate by a newly identified family of higher inositol polyphosphate kinases. *Curr. Biol.* **9**, 1323–1326.

27. York, S. J., Armbruster, B. N., Greenwell, P., Petes, T. D., & York, J. D. (2005) Inositol diphosphate signaling regulates telomere length. *J. Biol. Chem.* **280**, 4264–4269.

28. Laussmann, T., Eujen, R., Weisshuhn, C. M., Thiel, U., & Vogel, G. (1996) Structures of diphospho-myo-inositol pentakisphosphate and bisdiphospho-myo-inositol tetrakisphosphate from Dictyostelium resolved by NMR analysis. *Biochem. J.* **315**, 715–720.

29. Mulugu, S., Bai, W., Fridy, P. C., Bastidas, R. J., Otto, J. C., Dollins, D. E., Haystead, T. A., Ribeiro, A. A., & York, J. D. (2007) A conserved family of enzymes that phosphorylate inositol hexakisphosphate. *Science.* **316**, 106–109.

30. Choi, J. H., Williams, J., Cho, J., Falck, J. R., & Shears, S. B. (2007) Purification, sequencing, and molecular identification of a mammalian PP-InsP5 kinase that is activated when cells are exposed to hyperosmotic stress. *J. Biol. Chem.* **282**, 30763–30775.

31. Fridy, P. C., Otto, J. C., Dollins, D. E., & York, J. D. (2007) Cloning and characterization of two human VIP1-like inositol hexakisphosphate and diphosphoinositol pentakisphosphate kinases. *J. Biol. Chem.* **282**, 30754–30762.

32. Lin, H., Fridy, P. C., Ribeiro, A. A., Choi, J. H., Barma, D. K., Vogel, G., Falck, J. R., Shears, S. B., York, J. D., & Mayr, G. W. (2009) Structural

analysis and detection of biological inositol pyrophosphates reveal that the family of VIP/diphosphoinositol pentakisphosphate kinases are 1/3-kinases. *J. Biol. Chem,* **284**, 1863–1872.

33. De Vivo, M., Chen, J., Codina, J., & Iyengar, R. (1992) Enhanced phospholipase C stimulation and transformation in NIH-3T3 cells expressing Q209LGq-alpha-subunits. *J. Biol. Chem.* **267**, 18263–18266.

34. Otto, J. C., Kelly, P., Chiou, S. T., & York, J. D. (2007) Alterations in an inositol phosphate code through synergistic activation of a G protein and inositol phosphate kinases. *Proc. Natl. Acad. Sci. U S A.* **104**, 15653–15658.

35. Otto, J. C., Mulugu, S., Fridy, P. C., Chiou, S. T., Armbruster, B. N., Ribeiro, A. A., & York, J. D. (2007) Biochemical analysis of inositol phosphate kinases. *Meth. Enzymol.* **434**, 171–185.

36. Stevenson-Paulik, J., Chiou, S.-T., Frederick, J. P., dela Cruz, J., Seeds, A. M., Otto, J. C., & York, J. D. (2006) Inositol phosphate metabolomics: Merging genetic perturbation with modernized radiolabeling methods. *Methods.* **39**, 112–121.

37. Safranyl, S. T., Caffrey, J. J., Yang, X., Bembenek, M. E., Moyer, M. B., Burkhart, W. A., & Shears, S. B. (1998) A novel context for the 'MutT' module, a guardian of cell integrity, in a diphosphoinositol polyphosphate phosphohydrolase. *EMBO J.* **17**, 6599–6607.

38. Fujii, M., & York, J. D. (2005) A role for rat inositol polyphosphate kinases rIPK2 and rIPK1 in inositol pentakisphosphate and inositol hexakisphosphate production in Rat-1 Cells. *J. Biol. Chem.* **280**, 1156–1164.

Chapter 4

A Femtomole-Sensitivity Mass Assay for Inositol Hexakisphosphate

Andrew J. Letcher, Michael J. Schell, and Robin F. Irvine

Abstract

Inositol hexakisphosphate ($InsP_6$) is an important component of cells, and its mass levels are usually assayed by either (a) equilibrium labelling of cell cultures with radiolabelled inositol or (b) by a variety of mass assays of differing sensitivities and ambiguities. Here, we describe a mass assay for $InsP_6$ that is based on phosphorylating $InsP_6$ with $[^{32}P]$-ATP to $5\text{-}(PP)InsP_5$ using a recombinant *Giardia* $InsP_6$ kinase and quantification of the radiolabelled $5\text{-}[^{32}P](PP)InsP_5$ product by anion exchange HPLC with an internal $[^3H]\text{-}(PP)InsP_5$ standard. Interference with the enzyme reaction by other factors in the tissue extract is corrected for by assay of identical aliquots of tissue spiked with known amounts of $InsP_6$. This assay only measures $InsP_6$ (and not other inositol phosphates), and although it is simple in principle and requires no dedicated or specialised equipment, it is quite time-consuming. But the assay is unambiguous and is capable of quantifying accurately as little as 10 fmol of $InsP_6$ in a cell extract.

Key words: Inositol, Inositol phosphate, Inositol hexakisphosphate, Phytic acid, Phytate, Mass assay

1. Introduction

The "higher" (>4 phosphates) inositol phosphates are now the subject of intense interest in cell biology as an increasing body of evidence identifies the enzymes that synthesise them, and a diverse range of functions for an increasing number of them (see e.g., (1–3) for reviews, and (4–6) for more recent discoveries). Since inositol hexakisphosphate ($InsP_6$) was first found to be present in animal cells (7, 8) additional to its long-known function in plant tissues (9, 10), it has been the subject of particular interest. $InsP_6$ is present at (probably) 50–200 µM in most eukaryotic cells, though slime moulds apparently have levels nearer to 300 µM

Christopher J. Barker (ed.), *Inositol Phosphates and Lipids: Methods and Protocols*, Methods in Molecular Biology, vol. 645,
DOI 10.1007/978-1-60327-175-2_4, © Humana press, a part of Springer Science+Business Media, LLC 2010

(see (11) for references), which simplistically defies the chemical behaviour of the Mg^{2+} salt of $InsP_6$ in terms of its solubility limits (12).

The mass levels of $InsP_6$ are most usually assessed indirectly by equilibrium labelling with tritiated inositol, a method with its own interpretational problems (1) only suitable for stable cell cultures. The principal ways in which it has been directly assayed by mass include phosphate determination (13), metal dye detection coupled with HPLC (14, 15), or Dowex chromatography followed by dephosphorylation and mass spectroscopic quantification of myo-inositol in the higher inositol phosphate fraction (16). Each of these methods has strengths and weaknesses, and the weaknesses rule them out for some experiments that we plan to address issues such as: (a) how much $InsP_6$ in mammals is derived from exogenous sources (e.g., diet)? and (b) which metabolic route of endogenous synthesis is most important? For definitive answers to questions of this sort we need to assay $InsP_6$ (a) exactly, (b) unambiguously, and (c) to a very high level of sensitivity (so that if required we can set a very low limit on the levels of $InsP_6$ that may be present in a sample).

Using kinases in mass assays is something in which we have previously had some considerable success with inositol lipids – the kinase enzyme ensures specificity (and thus unambiguity), and sufficient enzyme plus high specific activity $[^{32}P]$-ATP can ensure high sensitivity. In this way, we have assayed $PtdIns4P$ (17, 18), $PtdIns$ (via PI-PLC and DAG kinase (17)), and $PtdIns5P$ (18, 19). Here, we describe an assay for $InsP_6$ based on a recombinant $InsP_6$ kinase, which uses HPLC to identify unambiguously and quantify the $5\text{-}(PP)InsP_5$ (henceforth referred to as $InsP_7$) thus generated (24).

2. Materials

2.1. Tissue Extract Preparation

1. Deionized water.
2. 50% (w/v) trichoroacetic acid (BDH).
3. Water-saturated Diethyl ether (Fisher Scientific).
4. 1:10 (v/v) saturated ammonia (Fisher Scientific).

2.2. Solutions for Preparation of Recombinant Giardia InsP₆ Kinase

1. Sterile salt solution: 0.17 M KH_2PO_4, 0.72 M K_2HPO_4 (both from Fisher Scientific). To make this, dissolve 2.31 g of KH_2PO_4 and 12.54 g of K_2HPO_4 in 90 mL of H_2O. After the salts have dissolved, adjust the volume to 100 mL and autoclave.
2. Glucose solution: 50 mL of 10% filter-sterilised glucose.

3. Terrific Broth with salts: 12 g Tryptone (Oxoid) + 24 g yeast extract (Formedium) + 4 mL glycerol (BDH) in 900 mL H_2O. Autoclave. Take 40 mL of this (Terific Broth) and add 5 mL sterile salt solution (Subheading 2.2, item 1) and 5 mL of sterile glucose (Subheading 2.2, item 2), maintaining sterile conditions.

4. Luria Broth: Dissolve 6.25 g of LB broth (supplier) in 250 mL H_2O and autoclave.

5. Lysis buffer: 50 mM Tris + 0.5 mM EDTA + 10 mM $MgCl_2$ adjusted to pH 8.0, and sterilized by filtration.

6. Sonication buffer: 10 mL of lysis buffer plus 100 µL of Calbiochem Protease Inhibitor Cocktail Set III (539134) + 10-µL Novagen Benzonase Nuclease (cat number 70746).

7. Dialysis buffer: 50 mM Tris, 1 mM EDTA buffered to pH 8 with NaOH.

8. Talon beads: Talon metal affinity resin (Clontech). Prepare these by spinning at $700 \times g$ in a bench centrifuge. Take off the supernatant using a pipette, removing the last traces with a narrow drawn-out tip. Add 10 mL of sonication buffer (above), shake and spin. Carry out a second wash and remove the supernatant.

9. Talon bead column buffer: 20 mM HEPES, 100 mM KCl, 5 mM imidazole, buffered to pH 7.0 with HCl.

10. Talon bead elution buffer: 20 mM HEPES, 100 mM KCl, 500 mM imidazole, buffered to pH 7.0 with HCl.

2.3. Mass Assay

1. Assay buffer 10×: 500 mM HEPES/NaOH pH 7.5, 1 M KCl, 50 mM $MgCl_2$, 10 mM EGTA, and 5 mg/mL Bovine serum albumin. Dilute 10× immediately before use.

2. 600 mM β-mercaptoethanol (BDH).

3. γ-[^{32}P]ATP (10 µCi) (NEN/Perkin Elmer).

4. Giardia InsP_6 kinase glycerol stock: The sequence of Giardia IP6K (Genbank accession #AY227443) was amplified from Giardia genomic DNA (the gene contains no introns), and then cloned into the hex-his bacterial vector pET43A, which fuses the solubility-enhancing bacterial protein NusA to the amino terminus of the fusion protein.

5. 0.1 M glucose (BDH).

6. Hexokinase (from *Saccharomyces cerevisiae* – Sigma).

7. Dowex Formate ion exchange resin: The Dowex 1X8-400, ion-exchange resin is obtained from ACROS ORGANICS (Fisher Scientific, UK), and is converted to the formate form by washing with 1 N NaOH followed by 1 M formic acid.

8. Econo Column (BioRad).

9. 0.4 M ammonium formate/0.1 M formic acid (both Sigma).

10. 0.8 M ammonium formate/0.1 M formic acid.

11. 2.5 M ammonium formate/0.1 M formic acid.

2.4. HPLC

1. Column: Whatman Partisil 10SAX WCS Analytical column, 4.6 mm × 250 mm (Cat No 4226-001) or a similar size column packed using the same material by Phenomenex.

2. 1.2 M Di-Ammonium hydrogen orthophosphate ($(NH_4)_2$ HPO_4) adjusted to pH 3.8 with orthophosphoric acid. (Fisher Scientific).

3. Scintillant: Ultima-Flo AP (Perkin-Elmer Cat 6013599).

2.5. Protein Assay

1. 8 M Urea (Sigma).

2. Decon ultrasonic bath.

3. BioRad Protein Assay reagent (Catalogue number 500-0006).

4. Bovine serum albumin (Sigma).

3. Methods

3.1. Tissue Extract Preparation

1. Tissue samples of 20–50 mg are weighed in a microtube and transferred to a 1 mL homogeniser with a ground glass pestle.

2. The sample is homogenised thoroughly in 180 μL of water and the homogenate is transferred to a 1.5 mL Eppendorf tube.

3. The homogeniser is washed out with a further 70 μL of water, which is pooled with the original homogenate.

4. To the homogenate is added 50 μL of 50% TCA to give a final concentration of 8.3%.

5. The sample is vortexed for 30 s and left on ice for 15 min.

6. The TCA precipitate is spun down in a microcentrifuge at 13 k ($13,000 \times g$) for 20 min at 4°C.

7. The supernatant is transferred to another Eppendorf tube and the TCA pellet is retained for protein estimation (see below).

8. The supernatant is given ten washes with 600 μL of water-saturated diethyl ether, then 15 μL of 1:10 (v/v) saturated ammonia is added and the tube is placed in a centrifugal vacuum dryer (Heto Maxi dry plus) for 5 min at 35°C followed by 30 min at 45°C to remove any remaining diethyl ether.

9. The remaining volume of liquid is measured using a micropipette, and the final volume is adjusted to 300 µL with water. The pH is checked at this time and is generally found to be about 6.

10. Using the original wet weight, the volume of the sample equivalent to 1 mg wet weight is taken for each assay.

3.2. Preparation of Giardia InsP₆ Kinase

1. Dispense 5 mL of Terrific Broth with salts and glucose (Subheading 2.2, item 3) into a 15-mL conical tube, and add 5 µL of Carbonicillin (25 mg/mL H_2O) plus 5 µL of Chloramphenicol (34 mg/mL H_2O). Spike with a few crystals of the Giardia IP6 kinase glycerol stock and grow overnight at 37°C.

2. Next day, add this to 250-mL Luria Broth (Subheading 2.2, item 4), and then add 0.25 mL of Carbonicillin (25 mg/mL H_2O) and 0.25 mL of Chloramphenicol (34 mg/mL H_2O).

3. Grow this culture at 37°C for 2–4 h until cloudy (OD 600 between 0.3 and 0.5) then transfer to an orbital shaker incubator (also at 16°C).

4. After another 6 h add IPTG (250 µL of 0.1 M prepared freshly in sterile water) to induce expression, and grow the culture overnight at 16°C on an orbital shaker.

5. Next day harvest the bacteria by spinning at 6,000×g for 15 min at 4°C and decant the supernatant. The pellets can be stored at this stage at –80°C.

6. Lysis: Add 10-mL sonication buffer (Subheading 2.2, item 6) to each pellet. Resuspend using a pipetman and transfer to 50-mL conical centrifuge tubes.

7. Sonication: This is done in a Soniprep 150 MSE Probe sonicator, keeping the sample on ice. Some ad hoc adjustments to conditions will need to made for other sonicators; that is, the ability of any sonicator to disrupt bacterial cells and release contents will need to be known, and if it is not, it will need to be explored before use (this can be assayed by gel analysis as described below). For our sonicator we use 3 × 30 s bursts at 12.5–13 with 30 s rests between (to prevent overheating, which will cause denaturation). The colour of the suspension should be a little darker if lysis has been successful.

8. Spin the sonicate for 30 min at 10,000×g at 4°C in Corex tubes in an MR 22i (Jouan) centrifuge. The pellet is a little fluid, so carefully remove the supernatant with a pipetman, and add this to 1 mL of Talon Metal Affinity beads (Subheading 2.2, item 8) in a 15-mL conical centrifuge tube.

9. Resuspend the sonication pellet in 10 mL of sonication buffer (Subheading 2.2, item 6) and retain a sample (as well as a

sample of the supernatant) for gel analysis, which will give an indication of the success of the sonication.

10. Place the conical centrifuge tube containing the sonication supernatant and the Talon beads on a rotating-wheel at room temperature for 60 min. Then spin the beads down in a bench centrifuge and remove the supernatant, saving some for a gel. Wash the beads by adding 10 mL of Talon column buffer (Subheading 2.2, item 9), place on the rotating wheel for 10 min, spin for 2 min, remove supernatant (saving some for analysis by gel run), and repeat this wash.

11. Finally, add 1 mL of column buffer (Subheading 2.2, item 9) to resuspend the beads, and transfer them to a Biorad mini-column. Allow this column to drain, and seal it. Add 1 mL of elution buffer (Subheading 2.2, item 10). Vortex and leave for several minutes. Remove the seal and collect the eluate in a 1.5-mL microtube (take 10-µL sample for a gel).

12. Repeat the elution with Talon bead elution buffer (Subheading 2.2, item 10) as above twice (retain 10 µL each time).

13. Pool the 3×1 mL eluates and place in a length of snake-skin dialysis tubing (Pierce).

    To do this, tie off one end, add a magnetic clip, and then add the 3×1 mL of Talon bead eluate. Tie-off the other end and add another clip (nonmagnetic). Suspend the sample for dialysis in 1 L of dialysis buffer (Subheading 2.2, item 7), add a disc-shaped stirring bar, and place on a stirrer in a cold-room. After 30 min or 1 h change the buffer; repeat twice more, and then leave stirring overnight. Change the buffer in the morning and leave for a further 2 h.

14. Recover the dialysate from the tubing, snap-freeze in 400-µL aliquots and store at –80°C.

15. When thawing an aliquot add an equal volume of glycerol and then store at –20°C.

3.3. Mass Assay

1. The standard assay uses seven tubes: Triplicate samples are taken of the tissue extract alone (equivalent of 1 mg wet weight, see Note 1) and of the tissue extract spiked with 10 pmol (see Note 2) of $InsP_6$ (see Note 3). The seventh tube is a water blank. These samples are made up to a final volume of 204 µL with water. To each is added:

   30 µL of 10× Assay Buffer,

   6 µL of 600 mM β-mercaptoethanol,

   10 µL of γ-[^{32}P]ATP (10 µCi) (see Note 4), and

   50 µL of Giardia $InsP6$ kinase glycerol stock (see Note 5).

2. Incubate for 60 min at 37°C.

3. Add $45\,\mu L$ 0.1 M glucose+$6\,\mu L$ hexokinase (20 units), and incubate at 37°C for a further 90 min. This step is to convert the remaining $\gamma[^{32}P]$-ATP to radiolabelled Glucose 6-phosphate+unlabelled ADP. Glucose 6-phosphate elutes much earlier from Dowex Formate than ATP or InsP_7, so the great majority of ^{32}P is removed before HPLC, and this (a) significantly decreases the background on the HPLC column, and (b) makes the assay safer by avoiding HPLC of highly radioactive samples.

4. Stop the reaction by adding 0.5 mL 0.4 M ammonium formate/0.1 M formic acid.

5. Load onto 0.5 mL of Dowex Formate in a BioRAd Econo Column (using a disc of Whatman Filter paper No 1 placed in top of the Dowex to stabilise the column).

6. Wash on with 0.5 mL followed by a further 4 mL of 0.4 M ammonium formate/0.1 M formic acid.

7. Wash with 10 mL 0.8 M ammonium formate/0.1 M formic acid to elute any remaining $\gamma^{32}P$-ATP.

8. Elute with 2×1 mL of 2.5 M ammonium formate/0.1 M formic acid.

9. The pooled sample is spiked with $[^3H]$-InsP_7 (prepared by phosphorylating $[^3H]$-InsP_6 with the Giardia InsP_6 kinase), and loaded onto the HPLC column.

3.4. HPLC

1. HPLC was carried out using the following elution protocol : Elution protocol: Solution A: Water, Solution B: 1.2 M Di-Ammonium hydrogen orthophosphate $((NH_4)_2HPO_4)$ pH 3.8. Flow rate: 1 mL/min.

2. Profile:

Time (min)	%B
0	0
5	0
30	78
75	78
76	100
81	100
82	0
102	0

3. 1 mL fractions are collected and counted for ^3H and ^{32}P by scintillation counting using an ammonium phosphate-tolerant scintillant, Ultima-Flo AP. Under these conditions, InsP_7 elutes approximately 60 min after injection.

3.5. Protein Assay

1. The TCA pellet is dissolved in 4 mL of 8 M Urea by sonication (Decon ultrasonic bath) and vortexing. The majority of the material dissolves but some solid particles may remain (see Note 6). If the tissue contains unusual amounts of fat this can remain in suspension and interfere with the assay.

2. Before assay, samples were diluted to give a final concentration of 6 M Urea, a concentration stated in the protocol to be compatible with the standard assay.

    Protein is assayed by the BioRad Protein assay, based on the Bradford method (20), with Bovine Serum Albumin as a standard.

4. Notes

1. The Km of the enzyme for $InsP_6$ is around 1 µM, so to keep the assay linear, amounts of $InsP_6$ greater than 100 pmol should be avoided.

2. The way of correcting for interfering factors in the cell extract by using a single "spiked" concentration assumes a linear effect of those factors on the assay. This correction is self-evidently more reliable if the "spike" of $InsP_6$ is of a similar magnitude to the $InsP_6$ present in the sample.

3. We have investigated whether $InsP_5$ isomers can be quantified by this technique, and the answer is that the assay is very much less sensitive. The major factor causing this is that the ^{32}P background in the region of the HPLC where the major products of $InsP_5$ phosphorylation by $InsP_6$ kinase elute, is very much higher than later (where $InsP_7$ elutes). Because of this we have not studied $InsP_5$s extensively, but our limited data suggest that the Giardia $InsP_6$ kinase is more specific for $InsP_6$ than some of its other eukaryote homologues (24). If this assay were to be used for $InsP_5$ isomers it will need significant adaptation.

4. The sensitivity of this assay rests on the signal-to-noise in the region of the HPLC where $InsP_7$ elutes. This is most influenced by the age of the $[^{32}P]$-ATP used. Over a period of a month or two when the $[^{32}P]$-ATP is usable the background increases significantly, so if very low levels of $InsP_6$ are to be measured, a fresh batch of $[^{32}P]$-ATP is best. It can be seen from the data in Fig. 1 that in that sample as little as 20–30 fmol of $InsP_6$ could be quantified by precise co-chromatography of the $[^{32}P]$-$InsP_7$ with the $[^3H]InsP_7$ standard. If a much smaller sample of the tissue had been taken for assay the "quench" caused by the extract would be less, and the

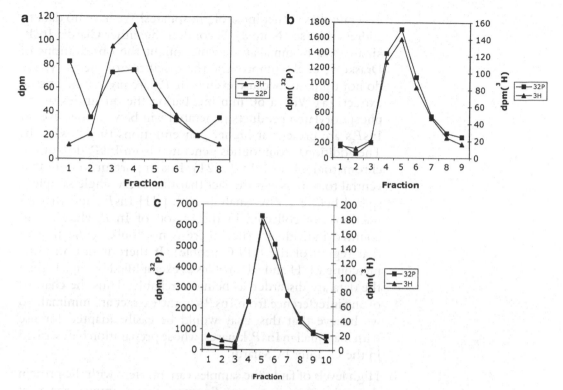

Fig. 1. Typical data from an experiment to measure InsP_6 in an extract of rat liver. For this assay, 50.5 mg wet weight of rat liver was extracted with 300 µL of 5%TCA as described. After removal of the TCA with diethyl-ether, the volume was adjusted to 300 µL, and 6×5 µL aliquots were taken for assay. Thus each aliquot contains the equivalent of 0.84 mg wet weight of liver. To three aliquots, 10 pmol of InsP_6 was added to correct for factors inhibiting the assay. Typical HPLC eluates are shown for individual examples of (**a**), a blank tube; (**b**), a liver sample; (**c**) a "spiked" liver sample. The time of elution of fraction 1 for each sample is about 57 min after injection. The [3H] data (*triangles*) show the elution of the [3H]-5*PP*-InsP_5 standard included in each sample, and the 32P data (*squares*) show the background (**a**) and amount of [32P]-5*PP*-InsP_5 generated in a control (**b**) and "spiked" (**c**) liver extract – note the different scales of the 32P axes. The data in (**b**) and (**c**) lead to a calculation of 3.95 pmol InsP_6 in sample B. This in turn leads to a calculation of 4.7 nmol InsP_6 per gramme wet weight tissue. Assuming 0.64-g water per gram of tissue (see refs. (22, 23) for derivation of this number) this gives a concentration of InsP_6 in hepatocytes (assuming uniform distribution and 100% recovery) of 7.3 µM.

sensitivity would be greater (indeed, with pure standards we can in principle detect about a femtomole of InsP_6 using fresh [32P]-ATP).

5. We have not explored other InsP_6 kinases, but see no reason why they should not be substituted for the Giardia enzyme. In that context, we should note that recently Draskovic et al. (21) have reported that mammalian InsP_6 kinases can, on prolonged incubation, produce additional inositol phosphates containing triphosphate groups. This additional complexity is especially pronounced when using InsP_5 as a substrate, and some of the compounds generated elute in the same region of an HPLC profile as InsP_7. These observations raise two questions: First, is the Giardia more or less likely to

generate these other inositol phosphates? We have no data to address this (see Note 3). Second, using either Giardia $InsP_6$ kinase or a mammalian enzyme, might the observations of Draskovic et al. compromise the specificity of the assay? We do not think so, because even if $InsP_5$s are present in a tissue extract: (a) With a 60 min incubation the quantities of tri-phosphorylated products generated will be very low, even if $InsP_5$s are present at higher concentrations than $InsP_6$. (b) The additional compounds generated from $InsP_5$ do not co-chromatograph with $InsP_7$ (21). It is important to note that central to our assay is the fact that that every single sample is spiked before HPLC analysis with [^{3}H]-$InsP_7$, and that *all* fractions are collected in the region of $InsP_7$ elution and counted individually (i.e., there is no "bulk" collection of that region of the HPLC profile). If there is not an *exact* matching of ^{3}H and ^{32}P profiles as exemplified by Fig. 1, then the data are discarded as being unreliable. Thus the chances of any interference from $InsP_5$s in the extract are minimal. So we believe that this assay would be easily adapted for use with mammalian $InsP_6$ kinases, whose preparation is described in the Chapter by Saiardi.

6. High levels of fat in the samples can interfere with the protein determination (see above). For many tissue samples, the most informative parameter is moles of $InsP_6$ per gramme fresh weight, because from that can be calculated the approximate concentration of $InsP_6$ in the cells (see legend to Fig. 1). But measuring the protein content is helpful to aid comparison with other studies that use per mg protein as a standard (cited in the Introduction), or when quantifying $InsP_6$ in, for example, cultured cells, where weighing the tissue extracted can be problematic.

Acknowledgments

This work was supported by the Wellcome Trust and the Royal Society. We thank Kevin Brindle for helpful discussions.

References

1. Irvine, R. F. and Schell, M. J. (2001) Back in the water: the return of the inositol phosphates. *Nat. Rev. Mol. Cell Biol.* **2**, 327–338

2. Shears, S. B. (2001) Assessing the functional omnipotence of inositol hexakisphophos-phate. *Cell. Signal.* **13**, 151–158.

3. York, J. D. (2006) Regulation of nuclear processes by inositol polyphosphates. *Biochim. Biophys. Acta.* **1761**, 552–559.

4. Mulugu, S., Bai, W., Fridy, P. C., Bastidas, R. J., Otto, J. C., Dollins, D. E., et al. (2007) A conserved family of enzymes that phosphorylate

inositol hexakisphosphate. *Science* **316**, 106–109.

5. Lee, Y. S., Mulugu, S., York, J. D. and O'Shea, E. K. (2007) Regulation of a cyclin-CDK-CDK inhibitor complex by inositol pyrophosphates. *Science* **316**, 109–112.

6. Illies, C., Gromada, J., Fiume, R., Leibiger, B., Yu, J., Juhl, K., et al. (2007) Requirement of inositol pyrophosphates for full exocytotic capacity in pancreatic beta cells. *Science* **318**, 1299–1302.

7. Heslop, J. P., Irvine, R. F., Tashjian, A. H., Jr. and Berridge, M. J. (1985) Inositol tetrakis- and pentakisphosphates in GH4 cells. *J. Exp. Biol.* **119**, 395–401.

8. Morgan, R. O., Chang, J. P. and Catt, K. J. (1987) Novel aspects of gonadotropin-releasing hormone action on inositol polyphosphate metabolism in cultured pituitary gonadotrophs. *J. Biol. Chem.* **262**, 1166–1171.

9. Posternak, S. (1919) Sur la synthese de l'ether hexaphosphorique de l'inosite avec le principe phospho-organique de reserve des plantes vertes. *C. R. Acad. Sci.* **169**, 138–140.

10. Raboy, V. (2003) myo-Inositol-1, 2, 3, 4, 5, 6-hexakisphosphate. *Phytochemistry* **64**, 1033–1043.

11. Torres, J., Dominguez, S., Cerda, M. F., Obal, G., Mederos, A., Irvine, R. F., et al. (2005) Solution behaviour of myo-inositol hexakisphosphate in the presence of multivalent cations. Prediction of a neutral pentamagnesium species under cytosolic/nuclear conditions. *J. Inorg. Biochem.* **99**, 828–840.

12. Veiga, N., Torres, J., Dominguez, S., Mederos, A., Irvine, R. F., Diaz, A., et al. (2006) The behaviour of myo-inositol hexakisphosphate in the presence of magnesium(II) and calcium(II): protein-free soluble InsP6 is limited to 49 microM under cytosolic/nuclear conditions. *J. Inorg. Biochem.* **100**, 1800–1810.

13. Stephens, L. R. and Irvine, R. F. (1990) Stepwise phosphorylation of myo-inositol leading to myo-inositol hexakisphosphate in Dictyostelium. *Nature* **346**, 580–583.

14. Mayr, G. W. (1988) A novel metal-dye detection system permits picomolar-range h.p.l.c. analysis of inositol polyphosphates from non-radioactively labelled cell or tissue specimens. *Biochem. J.* **254**, 585–591.

15. Lemtiri-Chlieh, F., MacRobbie, E. A. and Brearley, C. A. (2000) Inositol hexakisphos-phate is a physiological signal regulating the K$^+$-inward rectifying conductance in guard cells. *Proc Natl Acad Sci U.S.A.* **97**, 8687–8692.

16. Grases, F., Simonet, B. M., Prieto, R. M. and March, J. G. (2001) Variation of InsP$_4$, InsP$_5$ and InsP$_6$ levels in tissues and biological fluids depending on dietary phytate. *J. Nutr. Biochem.* **12**, 595–601.

17. Divecha, N., Banfic, H. and Irvine, R. F. (1991) The polyphosphoinositide cycle exists in the nuclei of Swiss 3 T3 cells under the control of a receptor (for IGF-I) in the plasma membrane, and stimulation of the cycle increases nuclear diacylglycerol and apparently induces translocation of protein kinase C to the nucleus. *EMBO J.* **10**, 3207–3214.

18. Clarke, J. H., Letcher, A. J., D'santos C. S., Halstead, J. R., Irvine, R. F. and Divecha, N. (2001) Inositol lipids are regulated during cell cycle progression in the nuclei of murine erythroleukaemia cells. *Biochem. J.* **357**, 905–910.

19. Morris, J. B., Hinchliffe, K. A., Ciruela, A., Letcher, A. J., and Irvine, R. F. (2000) Thrombin stimulation of platelets causes an increase in phosphatidylinositol 5-phosphate revealed by mass assay. *FEBS Lett.* **475**, 57–60.

20. Bradford, M. M. (1976) A rapid and sensitive method for the quantitation of microgram quantities of protein utilizing the principle of protein-dye binding. *Anal. Biochem.* **72**, 248–254.

21. Draskovic, P., Saiardi, A., Bhandari, R., Burton, A., Ilc, G., Kovacevic, M., et al. (2008) Inositol hexakisphosphate kinase products contain diphosphate and triphosphate groups. *Chem. Biol.* **15**, 274–286.

22. Reitzer, L. J., Wice, B. M. and Kennell, D. (1979) Evidence that glutamine, not sugar, is the major energy source for cultured HeLa cells. *J. Biol. Chem.* **254**, 2669–2676.

23. Brindle, K. M., Blackledge, M. J., Challiss, R. A. and Radda, G. K. (1989) 31P NMR magnetization-transfer measurements of ATP turnover during steady-state isometric muscle contraction in the rat hind limb in vivo. *Biochemistry* **28**, 4887–4893.

24. Letcher, A. J., Schell, M. J. and Irvine, R. F. (2008) Do mammals make all their own inositol hexakisphosphate? Biochem. J. **416**, 263–270.

Chapter 5

Synthesis of InsP_7 by the Inositol Hexakisphosphate Kinase 1 (IP6K1)

Cristina Azevedo, Adam Burton, Matthew Bennett, Sara Maria Nancy Onnebo, and Adolfo Saiardi

Abstract

Soluble inositol polyphosphates represent a variegate class of signalling molecules essential for the function of disparate cellular processes. Recently, the phytic acid derivate inositol pyrophosphate, InsP_7 (PP-IP$_5$ or IP$_7$) has been shown to pyro-phosphorylate proteins in a kinase independent way. To begin to understand the functional importance of this new phosphorylation mechanism, a source of cold and radiolabelled InsP_7 is indispensable. However, cold InsP_7 is expensive to buy, and labelled InsP_7 is not commercially available. Here we provide a protocol to synthesise and purify InsP_7 to a level of purity required for in vivo and in vitro experiments. We begin by purifying recombinant mouse inositol hexakisphosphate kinase (IP6K1) from *Escherichia coli*. With purified IP6K1, we produce cold InsP_7 and 5β[^{32}P] InsP_7 that we subsequently use in vitro experiments to phosphorylate proteins extracts from different species.

Key words: Inositol, Pyrophosphate, IP6K1, Phosphorylation, Protein purification, Enzymatic reaction

1. Introduction

Several phosphorylated derivatives of *myo*-inositol are present in eukaryotic cells (1, 2). Structurally, the most original are the inositol polyphosphates that contain one or two high energy pyrophosphate bonds (3). Of these inositol pyrophosphates, the best characterized are the diphosphoinositol pentakisphosphate (InsP_7 or PP-InsP_5) and the bis-diphosphoinositol tetrakisphosphate (InsP_8 or [PP]$_2$-InsP_4). InsP_7 and InsP_8 are present in all eukaryotic cells analyzed, from amoeba to mammalian neurons, and the enzymes responsible for their synthesis are highly conserved throughout evolution. In general, InsP_7 and InsP_8 levels are stable,

Christopher J. Barker (ed.), *Inositol Phosphates and Lipids: Methods and Protocols*, Methods in Molecular Biology, vol. 645, DOI 10.1007/978-1-60327-175-2_5, © Humana press, a part of Springer Science+Business Media, LLC 2010

representing around 5% of the concentration of their abundant precursor inositol hexakisphosphate (Phytic Acid or $InsP_6$) (4). However, $InsP_7$ and $InsP_8$ have a very rapid turnover; it has been calculated that 50% of the $InsP_6$ pool is converted to inositol pyrophosphates every hour (5). The dynamic nature of pyrophosphate turnover has led to the hypothesis that they may have a molecular switching activity and that in fact, many of the $InsP_6$ signalling roles may actually be carried out by inositol pyrophosphates. Several important cellular functions have been attributed to inositol pyrophosphates, from controlling telomere length to regulating chemotaxis (3). Furthermore, $InsP_7$ is important in modulating vesicular trafficking both in yeast, where its absence generates a fragmented vacuole (6), and in mammalian systems, where $InsP_7$ controls insulin secretion (7).

For some time, it was postulated that because of their high energy, inositol pyrophosphates could directly phosphorylate proteins. With the cloning of the IP6-Kinases (8), it was possible to confirm this hypothesis. IP6K1 was utilized to produce radiolabelled $InsP_7$ at the β-position of the pyrophosphate moiety ($5\beta[^{32}P]$ $InsP_7$), and phosphorylation of multiple proteins was detected in a kinase independent manner (9). Recently, it has been shown that this phosphorylation converts preexisting phosphoserine to pyrophosphoserine (10). Unfortunately, only one inositol pyrophosphate is currently commercially available, and therefore we use *Escherichia coli* expressing recombinant mouse IP6K1 to synthesize and purify cold and radiolabelled $InsP_7$. The synthesis and purification of these compounds is essential to further clarify the important physiological functions regulated by $InsP_7$.

2. Materials

Preparation of buffers and solutions described below (see Notes 1 and 2).

2.1. IP6K1 Affinity Tag Purification

2.1.1. Reagents

1. *E. coli* competent cells, BL21 (DE3).
2. 1 M Isopropylthiogalactoside (IPTG). Filter sterilized, dispense as 1 mL aliquots and store at –20°C.
3. Luria broth (Tryptone 10 g/L; Yeast Extract 5 g/L; NaCl, 10 g/L) agar and liquid medium.
4. Terrific broth (Tryptone 12 g/L; Yeast Extract 24 g/L; K_2HPO_4 9.4 g/L; KH_2PO_4 2.2 g/L; Glycerol 4 mL/L).
5. Selective antibiotics stock solutions: ampicillin (100 mg/L) and chloramphenicol (37.3 mg/L). Filter sterilize, dispense as 1 mL aliquots and store at –20°C.

6. Expression vectors: pGex-4T-2 (pGST-IP6K1) and pHisTrcA (pHis-IP6K1) (11). Available upon request.

7. Glutathione S-transferase (GST) agarose beads (Sigma-Aldrich G4510).

8. Talon Metal Affinity Resin (Clontech 635502).

9. Triton X-100.

10. 1 M Imidazole.

11. Protease inhibitor cocktail.

12. Any conventional protein concentration assay system.

13. Dialysis bags (Spectrum Labs 131267).

14. NuPAGE® LDS sample buffer 4× (Invitrogen NP0007).

15. NuPAGE® Novex 4–12% Bis-Tris gels 1.0 mm×10 well (Invitrogen NP0321).

16. Protein molecular weight marker.

17. Brilliant blue G.

18. His Lysis Buffer: 100 mM NaCl; 25 mM Tris–HCl pH 8.0. Store at 4°C.

19. His Wash Buffer I: 600 mM NaCl; 0.1% (v/v) Trition X-100; 25 mM Tris–HCl pH8.0. Store at 4°C.

20. His Wash Buffer II: 100 mM NaCl; 0.1% (v/v) Trition X-100; 25 mM Tris–HCl pH 7.4. Store at 4°C.

21. His Elution Buffer: 100 mM NaCl; 200 mM EDTA; 0.1% (v/v) Trition X-100; 25 mM Tris–HCl pH 7.4; add fresh 20 mM DTT. Store at 4°C.

22. His Dialysis Buffer: 100 mM NaCl; 1 mM MgSO$_4$; 1 mM DTT; 0.05% (v/v) CHAPS; 20 mM Tris–HCl pH 7.4. Store at 4°C.

23. GST Lysis Buffer: 20 mM Hepes pH 6.8; 100 mM NaCl; 1 mM EDTA; 1 mM EGTA; 0.1% (v/v) CHAPS; add fresh 5 mM DTT and 1/500 Protease inhibitor cocktail. Store at 4°C.

24. GST Wash Buffer: 50 mM Tris–HCl pH 7.4; 500 mM NaCl; 2 mM EDTA; 1% Triton X-100.

25. Coomassie blue staining solution: 0.5% (w/v) Brilliant blue G; 80% (v/v) Methanol; 20% (v/v) Acetic Acid.

26. Coomassie blue destaining solution: 25% (v/v) Isopropanol; 10% (v/v) Acetic Acid.

2.1.2. Equipment

1. Branson Sonifier 450 Sonicator or equivalent.

2. Head over tail rotator.

3. Chromatography mini Columns, 1.0×10 cm (BioRad 737-1011).

4. Chromatographic stand with Clamps.

5. XCell SureLock gel electrophoresis system (Invitrogen EI0001) or any conventional gel electrophoresis system.

2.2. Generation of Radiolabelled 5β[^{32}P] InsP$_7$

2.2.1. Reagents

1. Ultra-pure deionized water (resistivity of 18.2 megohm/cm).

2. Phosphoric acid (H_3PO_4).

3. Potassium carbonate (K_2CO_3).

4. Perchloric Acid.

5. Phytic Acid (InsP_6) (Sigma-Aldrich P8810).

6. γ[^{32}P] ATP (3,000 Ci/mmol).

7. ATP-Mg.

8. pH test strips range 4.0–10.0 (Sigma-Aldrich P4636).

9. Scintillation 6 mL mini-vials (Starstedt 73.680).

10. Ultima-Flo AP liquid scintillation cocktail (Perkin-Elmer 6013599).

11. 10× Hot Reaction Buffer: 200 mM Tris–HCl, pH 7.4; 500 mM NaCl; 60 mM MgSO$_4$; 0.5 mM ATP; 10 mM DTT and 0.5 mg/mL of BSA.

12. Neutralization Buffer: 1 M K$_2$CO$_3$ (stable at room temperature) and before use add 5 mM EDTA.

13. HPLC Buffers: The buffers are prepared using ultra pure water, filtered and degassed by vacuum filtration through inert 0.2 μM pore size membrane filter. Buffer A: 1 mM EDTA. Buffer B: 1 mM EDTA; 1.3 M (NH$_4$)$_2$HPO$_4$, pH 3.8 with H$_3$PO$_4$. Both buffers are stable at room temperature.

14. 1.5 M Triethylammonium bicarbonate: Mix in a volumetric flask 52 mL of triethylammonium bicarbonate in 198 mL of ice-cold water. Mix well and transfer approximately 50 mL of this solution into a beaker on ice; pH the solution with CO$_2$ to pH 8–8.3. Note that this step can take approximately 20 min and that toward the end the reaction goes faster.

15. 0.2 M Triethylammonium bicarbonate: Dilute with ultra pure water from the 1.5 M triethylammonium bicarbonate solution.

2.2.2. Equipment

1. pH meter.

2. Centrifuge evaporator.

3. Geiger counter.

4. Scintillation counter.

5. HPLC apparatus.

6. Partisphere SAX 4.6×125 column (Whatman 4621-0505).

7. Sep-Pak Accell Plus QMA cartridge (Waters WAT020545).

8. 1.5 mL screw cap tubes.

2.3. Generation of Cold InsP$_7$

2.3.1. Reagents

1. Creatine Phosphokinase (Sigma-Aldrich C3755).

2. Phosphocreatine (Sigma-Aldrich P7936).

3. Trace amount of tritium [^{3}H]-InsP$_6$ extracted from [^{3}H]-inositol yeast labelled cells (12), and purified according to Subheading 3.1.3 from step 4 onwards.

4. 5× Cold Reaction Buffer: 100 mM Tris–HCl, pH 7.4; 250 mM NaCl; 30 mM MgSO$_4$; 5 mM DTT and 0.2 mg/mL of BSA (see Note 2).

5. The other reagents are the same as for the generation of radiolabelled 5β[^{32}P] InsP$_7$.

2.3.2. Equipment

Same as for the generation of radiolabelled 5β[^{32}P] InsP$_7$.

2.4. InsP$_7$ Mediated Phosphorylation

2.4.1 Reagents

1. Phosphorylation Buffer I: 20 mM HEPES, pH 6.8; 1 mM EGTA; 1 mM EDTA, 0.1% (v/v) CHAPS; add fresh, 5 mM DTT, 5 mM NaF, 200 mg/mL PMSF and protease inhibitor cocktail. Aliquot and keep frozen.

2. Phosphorylation Buffer II: 20 mM Tris–HCl, pH 8.0; 50 mM NaCl, 6 mM MgCl$_2$; add fresh 1 mM DTT and 0.5 mM NaF.

2.4.2. Equipment

1. Autoradiography cassette.

2. X-ray film.

3. XCell SureLock gel electrophoresis system (Invitrogen EI0001) or any conventional gel electrophoresis system.

4. Optional: Western blot apparatus, Nitrocellulose or PVDF membrane.

3. Methods

Mammalian genomes possess three evolutionary conserved IP6Ks (IP6K1, IP6K2, and IP6K3) (13). In our experience, all three IP6Ks enzymes are expressed well in *E. coli*. However, IP6K3 is often present in inclusion bodies and is consequently the most difficult enzyme to purify out of the three. IP6K1 and 2 are expressed to similar levels both as His-tagged or GST-tagged entity, and no difference in stability or activity are detectable between these two enzymes. However, the nature of the tag affects the enzymatic capability of this class of enzymes. As shown in Fig. 1, GST-IP6K1 is able to produce InsP$_7$ generating only small amounts of InsP$_8$. In contrast, His-IP6K1 is able to easily convert InsP$_6$ to InsP$_7$, InsP$_8$, and even InsP$_9$. We take advantage of this difference in enzymatic ability to synthesize cold and hot (5β[^{32}P] InsP$_7$) InsP$_7$. For the synthesis of cold InsP$_7$ we use GST-IP6K1 and ATP

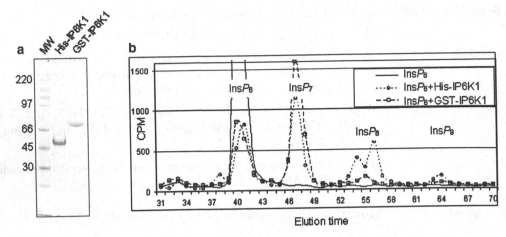

Fig. 1. Purification and activity of Tagged-IP6K1 proteins. (a) Visualization of purified proteins. 10 µl of purified GST-IP6K1 and His-IP6K1 proteins were analyzed with SDS-PAGE gel electrophoresis. Coomassie staining of the gel was performed to confirm the expected size of recombinant proteins and to assess the quality of the samples. The His-IP6K1 (53 kDa) sample is more that 95% pure as observed by a faint smear while the clean band (70 kDa) from the GST-IP6K1 sample indicate an almost pure sample. (b) HPLC analysis of inositol pyrophosphates. Purified GST-IP6K1 and His-IP6K1 were incubated in the presence of [^{3}H] InsP_6 and ATP to generate InsP_7 and analyzed by HPLC. GST-IP6K1 produced high levels of InsP_7 and close to zero amount of IP$_8$ and IP$_9$ while His-IP6K1 generated slightly lower amounts of InsP_7 as well as higher levels of InsP_7 and InsP_8.

recycling system (phosphocreatine, creatine phosphokinase) to drive the reaction forward. To synthesize 5β[^{32}P] InsP_7, we use instead His-IP6K1 without the ATP recycling system to avoid diluting the specific activity of γ[^{32}P]ATP with the phosphate coming from phosphocreatine. In this condition, the activity of His-IP6K1 only generates 5β[^{32}P]InsP_7.

3.1. IP6K1 Purification

3.1.1. Production of the E. coli

1. Transform 10 ng of pGST-IP6K1 or pHis-IP$_6$K1 into BL21 (DE3) competent cells and plate onto LB agar containing 100 µg/mL ampicilin, incubate overnight at 37°C.

2. Pick a single colony and grow in 5 mL of LB with 100 µg/mL ampicilin and 37.3 µg/mL chloramphenicol overnight at 37°C (see Note 4).

3. Dilute cells into 500 mL LB media with 100 µg/mL ampicilin and 37.3 µg/mL chloramphenicol and grow overnight at 37°C (in a 2,000-mL flask).

4. Next day, add 500 mL of prewarmed to 37°C LB media with 100 µg/mL ampicilin and grow for further 30 min (see Note 5).

5. Induce the expression by adding 1 M IPTG to a final concentration of 1 mM and incubate 3.5–4 h at 37°C with vigorous shaking.

6. Pellet the cells by centrifugation at 3,000×g for 15 min at 4°C.

7. Discard supernatant, add 40 mL of ice-cold water, transfer to 50 mL falcon tube, and spin as above.

8. Discard supernatant. At this point, cells can be frozen at −70°C until needed or proceed to the purification protocol accordingly.

3.1.2. His-IP6K1 Purification

1. Resuspend the pellet in 12 mL of ice-cold His Lysis buffer (see Notes 2, 3, and 6).

2. Lyse the cells on ice by sonication until the solution becomes translucent (Using the Branson Sonifier 450 instrument: four rounds of 20 s at setting 6) (see Note 7).

3. Pellet lysate by centrifugation at $10,000 \times g$ for 20 min at 4°C, remove supernatant to 14 mL falcon (retain 10 µl for SDS-PAGE analysis).

4. Add 1 mL of Talon Resin (previously equilibrated in His Lysis Buffer). Add Triton X-100 to a final concentration of 0.5% (v/v).

5. Incubate for 1 h at 4°C with head over tail rotation.

6. Transfer to a mini chromatographic column:
   – Allow the column to elute by gravity.

7. Wash as follows (between each step allow the column to elute by gravity):
   a. Close the tap and batch wash once with 7 mL of His Wash buffer I for 10 min.
   b. Add 7 mL of His Wash Buffer I.
   c. Add 7 mL of His Wash Buffer I with 5 mM Imidazole.
   d. Add 7 mL of His Wash Buffer II.

8. Elute as follows (collect 0.5 mL fractions; allow to flow by gravity):
   a. Add 0.5 mL of Elution Buffer and collect.
   b. Close tap, add 1 mL of Elution Buffer, and incubate 5 min with rotation and collect (repeat this step once more).
   c. Add 4 mL of Elution Buffer and collect.
   d. Note: Work fast and keep everything at 4°C.

9. To identify the fractions with the highest protein content determine protein concentration with appropriate method (e.g., Bio-Rad Protein assay).

10. Pool the fractions with highest protein concentration and transfer the sample to a dialysis bag. Dialyze as follows (keep all buffers at 4°C):
    a. Submerge the dialysis bag in 1 L of Dialysis Buffer containing 20% (v/v) glycerol and stir overnight at 4°C.

b. Remove buffer and add 0.5 L of Dialysis Buffer containing 40% (v/v) glycerol and stir 4 h at 4°C (repeat this step once more).

11. Remove sample from bag and store at −80°C until further use (Fig. 1) (see Note 8).

12. Regenerate the column as follows (in between steps allow the washing to flow by gravity):

a. 5 mL of His Elution Buffer.

b. 15 mL of H_2O.

c. 15 mL of 50 mM $CoCl_2$.

d. 8 mL of H_2O.

e. 4 mL of 300 mM NaCl.

f. 4 mL of H_2O.

13. Leave the resin sealed in 20% (v/v) ethanol at room temperature.

3.1.3. GST-IP6K1 Purification

1. Resuspend pellet in 12 mL of ice-cold GST Lysis Buffer and set aside 10 μl for analysis by SDS-PAGE (see Note 3).

2. Lyse the cells by sonication until the solution becomes translucent.

3. Pellet lysate by centrifugation at $10,000 \times g$ for 20 min at 4°C and transfer the supernatant to a 14-mL falcon (retain 10 μl for SDS-PAGE analysis).

4. Add 0.5 mL of Glutathione beads (preequilibrated twice in GST Lysis Buffer). Add Triton X-100 to a final concentration of 1% (v/v).

5. Incubate at 4°C with head over tail rotation for 2 h.

6. Wash beads with 10 mL of GST Wash Buffer for 5 min. Spin for 2 min at 500 rpm, and carefully remove supernatant in order not to disturb the beads (repeat this step three more times).

7. Equilibrate the beads by washing with 1 mL of GST Lysis Buffer (Fig. 1) (see Note 8).

3.1.4. In Vitro IP6K1 Assay and Purification of Radiolabel InsP7 (5β[32P] InsP7) (9, 14)

1. Incubate overnight at 37°C five reactions containing in 1× Hot Reaction Buffer, 100 ng of His-IP6K1, 400 μM $InsP_6$, and 60 μCi of γ[^{32}P]ATP (in a total volume of 20 μl) (see Note 9).

2. Pool the five reactions together.

3. Quench the reaction by adding the same volume (100 μl) of ice cold 0.6 M perchloric acid.

4. Neutralize with the appropriate volume of neutralization solution. As a starting point, use half of the total volume. After neutralization, the sample pH should be between 6 and 8. Use pH test strips to measure the sample pH. If the pH is not

within the referred interval, add a small portion of perchloric acid or neutralization solution accordingly. Add neutralization solution slowly and a little at time as the CO_2 generated tends to bubble over.

5. Leave the tubes loosely capped on ice for at least 2 h. Gently flick every 30 min.

6. Spin in a refrigerated centrifuge at maximum speed for 10 min.

7. Carefully transfer the supernatant to a fresh tube without touching the pellet.

8. The synthesized $5\beta[^{32}P]$ InsP$_7$ is then purified by HPLC using a Partisphere SAX column. Prepare the HPLC for sample injection washing the column with buffer A for about 30 min taking care to gradually increase the flow rate from 0.1 to 1 mL/min (see Note 10).

9. Inject the sample and run the HPLC according to the following method 0–3 min, 0% B; 3–8 min, 0–30% B; 8–48 min, 100% B, setting the fraction collector to collect 1 mL fractions.

10. Once the run is finished, rapidly check the radioactivity of the eluted fractions using a Geiger counter. Using the indicated chromatographic procedure, unincorporated ATP should be eluted between fraction 11–16, whereas newly synthesized $5\beta[^{32}P]$ InsP$_7$ will elute around fraction 34–37.

11. Wash the HPLC system with buffer A after each use and leave the instrument running overnight at 0.1 mL/min.

12. Pool the samples containing $5\beta[^{32}P]$ InsP$_7$.

13. Neutralize with 50% (v/v) ammonia (usually 200 µl/mL). Test neutralization with pH strip as before.

14. Dilute the neutralized fractions 17-fold with cold water (3 mL in 50 mL if pooled three fractions).

15. Prewash a Sep-Pak Accell Plus QMA cartridge with 10 mL of H_2O applying steady pressure.

16. Load (at 2 mL/min) the diluted sample onto the cartridge with a 50-mL syringe.

17. Wash the column with 4 mL of ice-cold 0.2 M triethylammonium bicarbonate.

18. Elute the radioisotope with 4 mL of 1.5 M triethylammonium bicarbonate and collect in three 1.5 mL centrifuge tubes. Check the radioactivity with the Geiger counter; the first two elution fractions should contain more that 90% of radioactivity; throw away the third tube.

19. Vacuum dry the samples until a volume of approximately 20 µl, making sure the samples are not completely dry; otherwise, it will be difficult to resuspend all the InsP$_7$ purified.

20. Determine the radioactivity counting 1 µl of the sample using a scintillation counter. Usually, the recovery is 70–100×10^6 CPM of $5\beta[^{32}P]$ $InsP_7$.

3.1.5. In Vitro IP6K1 Assay and Purification of Cold InsP₇

Take special attention throughout the following procedure (see Notes 9 and 10).

1. Prepare five (or more) reactions in 1× Cold Reaction Buffer containing 100 ng of GST-IP6K1, 400 µM $InsP_6$, 1 mM ATP, 10 mM Phosphocreatine, 40 U Creatine Phosphokinase, and 2×10^4 CPM of $[^3H]$-$InsP_6$ (in a total volume of 25 µl) (see Note 9).

2. Prepare the sample and run the HPLC as reported for the purification of radiolabelled $InsP_7$ (points 2–9).

3. Collect and pool the fractions that contain the newly synthesized $InsP_7$ and desalt them using the protocol described for the purification of radiolabelled $InsP_7$ (points 13–19).

4. Determine the radioactivity by counting 1 µl of the sample using a scintillation counter. Knowing the initial concentration of the cold $InsP_6$ used and the percent of $[^3H]$-$InsP_6$ recovered as $[^3H]$-$InsP_7$, the concentration of $InsP_7$ can be calculated.

3.1.6. InsP₇ Mediated Protein Phosphorylation

You can perform this assay on protein extracts, on purified proteins, or on protein extracts overexpressing your favorite protein that is either tagged or to which you have a specific antibody

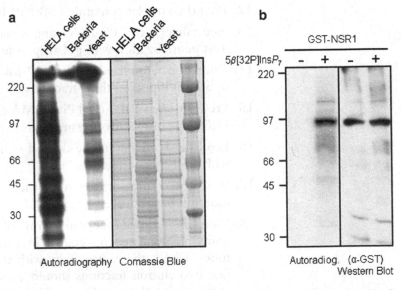

Fig. 2. $InsP_7$ mediated phosphorylation of protein extracts. (a) Autoradiography of $InsP_7$ mediated phosphorylation of protein extracts from yeast, bacteria, and HeLa cells (*left*) and Coomassie blue staining of the respective SDS-PAGE gel (*right*). (b) $InsP_7$ mediated phosphorylation of yeast protein extracts expressing GST-NSR1, the extract was run on an SDS-PAGE, transferred to a PDVF membrane and autoradiographed (*left*), subsequently the membrane was blotted with anti GST antibody (*right*).

against (Fig. 2). If using purified protein, it is not required to elute it from the beads; in fact, having the protein on the beads is an advantage since it can be equilibrated directly in the appropriate buffer. If using purified protein, proceed to point three of this protocol (see Note 9).

1. Extract protein in ice-cold Phosphorylation Buffer I (see Note 3).

2. Centrifuge cell lysate for 20 min at $15,000 \times g$ at 4°C, remove supernatant to fresh tube.

3. Incubate 10–20 µg of protein extracts (or 0.1–1 µg of purified protein) with Phosphorylation Buffer II in the presence of 10^6 CPM of 5β[^{32}P] InsP_7 in a total volume of 20 µl, at 37°C for 10 min.

4. Add 4× sample buffer, boil 5 min at 95°C. Keep on ice until needed.

5. Run samples on SDS-PAGE gel.

6. Depending on the experiment gels can either be stained with coomassie blue stain and dried or transferred to nitrocellulose or PVDF membrane (see Notes 11 and 12).

7. Expose dried gel or the membrane to X-ray film for at least 24 h before developing (keep cassette at –70°C for stronger signal; expose for longer time – up to 4 days, if the signal is weak).

8. Once film is developed, the membrane can be subjected to immunoblotting according to standard procedures in order to detect the presence of your favorite protein (if the signal from the radioactive labelling is strong, it is advisable to allow it to decay for 1–2 weeks before proceeding to immunoblotting).

4. Notes

1. Several of the reagents proposed throughout these protocols can be substituted by equivalent ones from other commercially available sources. For instance, we use the precast gel system NuPAGE from Invitrogen for convenience and reliability, but this system is by no means essential; you can pour your own gels or use other precast gels system as preferred in your laboratory. Moreover, we only reference the source of the reagents that might be less common in the laboratories.

2. All buffers should be prepared with ultra-pure deionized water (resistivity of 18.2 megohm/cm) and kept at 4°C, unless otherwise stated.

3. All the protein work should be done at 4°C or on ice.

4. It is advisable to do a small pilot screening for fusion protein expression before proceeding to the large scale purification. This is done by growing individually 2–4 colonies expressing GST-IP6K1 or His-IP6K1 in 3 mL until $OD_{600} = 0.6$–0.8 at 37°C. Before inducing with 1 mM IPTG, remove 20 µl of culture (non-induced control). After 1–2 h of IPTG induction, pellet cells, resuspend 300 µl in GST Lysis Buffer or His Lysis buffer accordingly and sonicate. Pellet debris and transfer supernatant to fresh tube, add 4× sample buffer, boil and run 20 µl on SDS-PAGE. Stain the gel with coomassie blue to visualize the induction of the GST-IP6K1 (73 kDa) or His-IP6K1 (54 kDa) fusion at the expected size. Make glycerol stock of a colony with good level of expression.

5. During bacterial growth ensure adequate aeration, do not fill the flasks more than 25% of their volume and use a permeable lid, making sure sterility is kept inside the flask.

6. It is advisable to retain small samples at key steps of the purification procedure to analyze by SDS-PAGE in case the yield is not as expected. This way it is easier to identify at which point the protein was lost.

7. During sonication, cell lysis is evident by a change in the extract from cloudy to translucent. Try to keep the probe in the centre of the tube avoiding frothing as this may denature the protein. Also, avoid oversonication as this increases purification of contaminant proteins.

8. His-IP6K1 yield may be determined by standard chromogenic procedures (like BCA or Bradford) whereas GST-IP6K1 yield, because we do not elute the protein from the beads, can be determined empirically by direct comparison with BSA standards on a coomassie blue stained SDS-PAGE gel.

9. Take appropriate precautions when handling radioactivity and always follow your laboratory good practice rules throughout the experiments. To prevent radioactive contamination through leakage, we use screw cap tubes instead of the common snap cap.

10. To secure accurate and reproducible HPLC runs, it is vital that the pH and concentration of buffer B are accurate, since small changes in these parameters can affect the elution profile. When making a new batch of buffer B, it is always advisable to run radioactive standards to check the elution profile. The reproducibility of subsequently HPLC runs is of crucial importance for the purification of cold $InsP_7$, because the fractions containing trace amounts of tritium-$InsP_7$ cannot be detected by Geiger or scintillation counting.

11. To stain the gel, immerse it in Coomassie Blue staining solution for about 30 min at room temperature with shaking.

Rinse in water and add excess of coomassie de-staining solution until the bands are clear. Remember that the gel is radioactive, and therefore appropriate measures must be taken in handling the gel and in disposing the solutions.

12. If transferring the gel to a membrane follow standard western blot procedures.

Acknowledgements

This work is supported by MRC funding of the Cell Biology Unit and by EC through an IRG (014827).

References

1. Irvine R.F. and Schell M.J. (2001) Back in the water: the return of the inositol phosphates. Nat Rev Mol Cell Biol. **2**: 327–338.

2. Shears S.B. (2004) How versatile are inositol phosphate kinases? Biochem J. **377**: 265–280.

3. Bennett M., Onnebo S.M., Azevedo C. and Saiardi A. (2006) Inositol pyrophosphates: metabolism and signaling. Cell Mol Life Sci. **63**: 552–564.

4. Raboy V. (2003) myo-Inositol,2,3,4,5,6-hexakisphosphate. Phytochemistry. **64**: 1033–1043.

5. Glennon M.C. and Shears S.B. (1993) Turnover of inositol pentakisphosphates, inositol hexakisphosphate and diphospho-inositol polyphosphates in primary cultured hepatocytes. Biochem J. **293**: 583–590.

6. Saiardi A., Sciambi C., McCaffery J.M., Wendland B. and Snyder S.H. (2002) Inositol pyrophosphates regulate endocytic trafficking. Proc Natl Acad Sci U S A. **99**: 14206–14211.

7. Illies C., Gromada J., Fiume R., Leibiger B., Yu J., Juhl K., et al. (2007) Requirement of inositol pyrophosphates for full exocytotic capacity in pancreatic beta cells. Science. **318**: 1299–1302.

8. Saiardi A., Erdjument-Bromage H., Snowman A.M., Tempst P. and Snyder S.H. (1999) Synthesis of diphosphoinositol pentakisphosphate by a newly identified family of higher inositol polyphosphate kinases. Curr Biol. **9**: 1323–1326.

9. Saiardi A., Bhandari R., Resnick A.C., Snowman A.M. and Snyder S.H. (2004) Phosphorylation of proteins by inositol pyrophosphates. Science. **306**: 2101–2105.

10. Bhandari R., Saiardi A., Ahmadibeni Y., Snowman A.M., Resnick A.C., Kristiansen T.Z., et al. (2007) Protein pyrophosphorylation by inositol pyrophosphates is a posttranslational event. Proc Natl Acad Sci U S A. **104**: 15305–15310.

11. Saiardi A., Caffrey J.J., Snyder S.H. and Shears S.B. (2000) The inositol hexakisphosphate kinase family. Catalytic flexibility and function in yeast vacuole biogenesis. J Biol Chem. **275**: 24686–24692.

12. Azevedo C. and Saiardi A. (2006) Extraction and analysis of soluble inositol polyphosphates from yeast. Nat Protoc. **1**: 2416–2422.

13. Saiardi A., Nagata E., Luo H.R., Snowman A.M. and Snyder S.H. (2001) Identification and characterization of a novel inositol hexakisphosphate kinase. J Biol Chem. **276**: 39179–39185.

14. Menniti F.S., Miller R.N., Putney J.W., Jr. and Shears S.B. (1993) Turnover of inositol polyphosphate pyrophosphates in pancreatoma cells. J Biol Chem. **268**: 3850–3856.

Chapter 6

Protein Pyrophosphorylation by Diphosphoinositol Pentakisphosphate (InsP_7)

J. Kent Werner Jr., Traci Speed, and Rashna Bhandari

Abstract

Diphosphoinositol polyphosphates, also known as inositol pyrophosphates, are a family of water soluble inositol phosphates that possess diphosphate or pyrophosphate moieties. In the presence of divalent cations such as Mg^{2+}, the "high energy" β phosphate can be transferred from the inositol pyrophosphates, InsP_7 and InsP_8, to prephosphorylated serine residues on proteins, to form pyrophosphoserine. This chapter provides detailed methods to identify proteins that are substrates for pyrophosphorylation by InsP_7, conduct phosphorylation assays on purified protein, and detect protein pyrophosphorylation.

Key words: Inositol pyrophosphate, Diphosphoinositol pentakisphosphate, Pyrophosphoserine, Pyrophosphorylation, Phosphorylation

1. Introduction

Diphosphoinositol polyphosphates, or inositol pyrophosphates, including InsP_7 and InsP_8, participate in a variety of cellular processes, including DNA recombination, telomere length maintenance, stress response, phosphate homeostasis, vesicle trafficking, apoptosis, insulin secretion, and spermatogenesis (1–4). Recent studies demonstrate that inositol pyrophosphates bring about these diverse effects via two basic molecular mechanisms (a) protein binding and (b) protein pyrophosphorylation. In *Dictyostelium discoideum*, 5-PP-InsP_5, an isomer of InsP_7, regulates chemotaxis by displacing the binding of PtdIns(3,4,5)P_3 to pleckstrin homology (PH) domain-containing proteins (5). The other InsP_7 isomer, 1/3-PP-InsP_5, specifically binds to the *S. cerevisiae* cyclin-dependent kinase inhibitor protein, PHO81, thereby controlling phosphate homeostasis (6).

Christopher J. Barker (ed.), *Inositol Phosphates and Lipids: Methods and Protocols*, Methods in Molecular Biology, vol. 645, DOI 10.1007/978-1-60327-175-2_6, © Humana press, a part of Springer Science+Business Media, LLC 2010

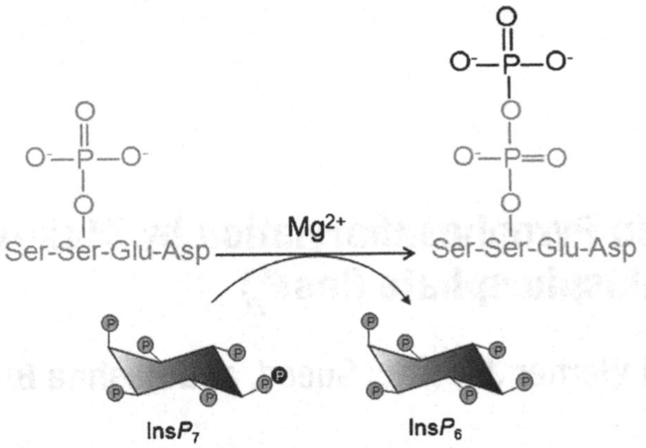

Fig. 1. Protein pyrophosphorylation. Inositol pyrophosphates such as InsP_7 transfer their β phosphate group to a prephosphorylated serine residue to generate pyrophosphoserine. Pyrophosphorylation occurs in the presence of divalent magnesium ions, and only on serine residues present in "acidic serine" motifs.

In 2004, it was discovered that the β phosphate moiety of InsP_7 can be transferred to serine residues in eukaryotic proteins (7). Subsequent studies demonstrated that InsP_7 can only phosphorylate prephosphorylated serine residues to form pyrophosphoserine (Fig. 1) (8). In vitro, pyrophosphorylation is an enzyme-independent reaction, requiring only the inositol pyrophosphate donor, the prephosphorylated protein acceptor, and divalent cations such as Mg^{2+}. Pyrophosphorylation can be brought about by all inositol pyrophosphates, including 5-PP-InsP_5, 1/3PP-InsP_5, and InsP_8. The acceptor serine residues are prephosphorylated by a protein kinase, usually CK1 or CK2, and occur in acidic serine sequence motifs, i.e., a stretch of two or more serine residues interspersed with Glu and/or Asp residues. Such sequences occur commonly throughout the proteome of all eukaryotic organisms. The β phosphate moiety in pyrophosphoserine is cleaved in the presence of acid but is insensitive to phosphatases such as lambda protein phosphatase and protein phosphatase-1, which rapidly dephosphorylate phosphoserine residues (8).

Since acidic serine sequence motifs occur frequently in eukaryotic proteins, it is likely that proteins of interest to individual biologists undergo inositol pyrophosphate-mediated serine pyrophosphorylation. Therefore, this chapter outlines methods to identify serine residues that may be pyrophosphorylated, carry out pyrophosphorylation assays on purified substrate protein using 5β[^{32}P]InsP_7, and confirm pyrophosphorylation by testing the biochemical properties of pyrophosphoserine, namely phosphatase insensitivity and acid lability, that distinguish it from phosphoserine.

2. Materials

See Note 1 before preparing solutions and media described throughout this section.

2.1. Identification of Pyrophosphorylation Sites/Sequence Motifs on Proteins

1. The sequence of your protein of interest can be downloaded from protein sequence databases, such as National Center for Biotechnology Information (NCBI, www.ncbi.nlm.nih.gov/) or Universal Protein Knowledgebase (UniProt, http://www.uniprot.org).

2. One of several available online pattern matching algorithms can be used, such as PatMatch (9).

2.2. Overexpression and Purification of a GST-Fusion Protein in Yeast

2.2.1. Extract from Yeast Expressing a GST-Fusion Protein

1. Incubator that can maintain 30°C.

2. Shaker incubator that can maintain 220–250 rpm and 30°C.

3. Spectrophotometer to measure cell density at 600 nm.

4. Centrifuge with swing bucket rotor that can hold 15 mL conical tubes.

5. Refrigerated centrifuge to hold 1.5–2 mL tubes.

6. Frozen EZ Yeast Transformation Kit (Zymo Research, T2001).

7. SC-URA + Dex (Synthetic complete medium without uracil with dextrose as the carbon source): Yeast Nitrogen Base with ammonium sulfate, without amino acids (Difco, 291940) 6.7 g/L, Yeast synthetic dropout medium supplements without uracil (Sigma-Aldrich, Y1501) 1.92 g/L, Dextrose (Difco, 215530) 20 g/L. Autoclave or filter-sterilize. To pour plates, add 1.6% agar and autoclave.

8. SC-URA + Gal (Synthetic complete medium without uracil with galactose as the carbon source): Yeast Nitrogen Base with ammonium sulfate, without amino acids (Difco, 291940) 6.7 g/L, Yeast synthetic dropout medium supplements without uracil (Sigma-Aldrich, Y1501) 1.92 g/L, D-Galactose (Difco, 216310) 20 g/L. Autoclave or filter-sterilize.

9. Lysis Buffer: 20 mM HEPES, pH 6.8; 1 mM EDTA, 1 mM EGTA, 0.1% CHAPS, 5 mM DTT, Protease Inhibitor cocktail for yeast extracts (Sigma-Aldrich, P8465).

10. Glass beads (425–600 µm).

11. Bradford protein estimation reagent or any other protein concentration assay system.

2.2.2. Purification of a GST-Fusion Protein from Yeast

1–5. Same as in Subheading 2.2.1.

6. Head over tail rotation apparatus.

7. Mini-gel apparatus for SDS-PAGE.

8. Cold room or 4°C cabinet.

9. GST Purification Buffer A: 20 mM HEPES pH 6.8, 100 mM NaCl, 2 mM EDTA, 5 mM DTT, Protease inhibitor cocktail for yeast extracts (Sigma-Aldrich, P8215).

10. GST Purification Buffer B: Buffer A + 1% Triton X-100.

11. GST Purification Buffer C: 20 mM HEPES, pH 6.8, 500 mM NaCl, 2 mM EDTA, 1% Triton X-100.

12. Triton X-100 10% (v/v).

13. Glutathione-agarose beads (Sigma-Aldrich, G4510). Swell beads in water and resuspend in water at 1:1 (v/v) ratio, according to the manufacturers' instructions.

14. Preparation of glutathione-agarose beads: Pipette 1 mL of glutathione-agarose beads from a 1:1 suspension (see Note 2) into a 15 mL conical tube. Wash the beads in GST Purification Buffer B (two washes of 3 mL each). For each wash, add Buffer B, mix the tube by inverting, and centrifuge at $1,000 \times g$ for 3 min at 4°C. Resuspend the beads in Buffer B to make a 1:1 suspension.

15. Phosphate buffered saline (PBS): 137 mM NaCl, 2.7 mM KCl, 10 mM Sodium phosphate dibasic, 2 mM Potassium phosphate monobasic, pH 7.4.

16. Tris-Glycine-SDS-gel for PAGE, cast and run as per standard protocols.

17. EZBlue Gel Staining Reagent (Sigma-Aldrich, G1041) or any Coomassie Brilliant Blue R or G based protein stain.

2.3. Prephosphorylation of a Protein Expressed in E. coli

1. Protein kinase CK1 (NEB, P6030) or CK2 (NEB, P6010) supplied with 10× kinase buffer, or any purified CK1 or CK2 (holoenzyme or alpha subunit) available commercially.

2. Mg-ATP.

3. Phosphate buffered saline (PBS) – see above.

4. 1× Phosphorylation Buffer: 25 mM Tris–HCl pH 7.4, 50 mM NaCl, 1 mM DTT, 6 mM $MgCl_2$.

2.4. Phosphorylation of Proteins by 5β[^{32}P] InsP$_7$

1. XCell SureLock gel electrophoresis system (Invitrogen EI0001), or any equivalent mini-gel system.

2. Western Blot apparatus, such as Mini Trans-Blot Electrophoretic Transfer Cell (Bio-Rad, 170-3930), or equivalent wet or semi-dry transfer apparatus.

3. PVDF membrane for protein transfer.

4. Heating block or water bath.

5. Phosphorimager or film and cassettes for autoradiography.

6. Perspex/plexiglass shielding to protect users from beta particles emitted by ^{32}P.

7. 5β[^{32}P]InsP_7 prepared as described in Chapter 5, Subheading 3.1.3.

8. Mg-ATP.

9. 10× Phosphorylation Buffer: 250 mM Tris–HCl pH 7.4, 500 mM NaCl, 10 mM DTT, 60 mM MgCl$_2$.

10. LDS sample buffer 4× (Invitrogen, NP0007).

11. NuPAGE® Novex 4–12% Bis–Tris gels 1.0 mm×10 well (Invitrogen, NP0321).

12. See Blue^R Plus2 (Invitrogen, LC5925) prestained protein standard, or any other protein molecular weights standard.

13. Antiglutathione-S-Transferase (GST)-Peroxidase Conjugate antibody produced in rabbit (Sigma-Aldrich, A7340).

2.5. Testing the Biochemical Properties of InsP_7-Phosphorylated Proteins

1. Heating blocks or water baths at 30, 37, and 95°C.

2. XCell SureLock gel electrophoresis system (Invitrogen EI0001), or any equivalent mini-gel system.

3. Gel dryer.

4. Phosphorimager or film and cassettes for autoradiography.

5. Mg-ATP.

6. γ[^{32}P]ATP, 3,000–6,000 Ci/mmole.

7. Protein kinase CK2 (NEB, P6010), or any purified CK2 (holoenzyme or alpha subunit) available commercially.

8. EDTA (500 mM, pH 8.0).

9. 5β[^{32}P]InsP_7 prepared as described in Chapter 5, Subheading 3.1.3.

10. 10× Phosphorylation Buffer – see Subheading 2.4.

11. PBS.

12. 6 M Hydrochloric acid (HCl), prepared from stock of concentrated acid.

12. LDS sample buffer 4× (Invitrogen, NP0007).

13. NuPAGE® Novex 4–12% Bis–Tris gels 1.0 mm×10 well (Invitrogen, NP0321).

14. EZBlue Gel Staining Reagent (Sigma-Aldrich, G1041) or any standard Coomassie Brilliant Blue R or G based protein stain.

15. Lambda protein phosphatase supplied with 10× Phosphatase Buffer (NEB, P0753).

3. Methods

The initial analysis of InsP_7-phosphorylated proteins by partial purification and mass spectrometry led to identification of the *S. cerevisiae* proteins NSR1 and YGR130c as InsP_7 substrates (7).

Deletion and mutagenesis analyses revealed that $InsP_7$ phosphorylates serine residues in an "acidic serine" sequence motif (multiple Ser residues interspersed with Glu/Asp residues), in both these proteins. Subsequently, novel $InsP_7$ substrates were identified by scanning proteins for acidic serine sequence motifs, and the yeast protein SRP40, and mammalian proteins TCOF1, Nopp140, and AP3β3 were found to be phosphorylated by $InsP_7$ (7). The methods outlined below describe the identification of an $InsP_7$ substrate protein by sequence analysis, its expression in a eukaryotic host, and phosphorylation by radiolabelled $InsP_7$. An alternative method is described if the protein is expressed in *E. coli*, wherein the prephosphorylation "priming" step is carried out using the protein kinase CK2. After phosphorylation of a purified protein by $InsP_7$, it can be examined for stability of the added phosphate in the presence of hydrochloric acid or lambda protein phosphatase, to confirm that the protein is pyrophosphorylated.

3.1. Identification of Potential Pyrophosphorylation Sites on Proteins

1. Obtain the amino acid sequence of your protein(s) of interest from NCBI or any other sequence database.

2. Scan the protein sequence manually, by eye, to identify the pyrophosphorylation consensus site i.e., two or more Ser residues interspersed with, or flanked by acidic residues, Glu or Asp.

3. Alternatively, use a sequence scanning algorithm to search for an acidic serine sequence motif in a protein sequence database. For example, the loosely defined consensus site for $InsP_7$-mediated phosphorylation, $[E/D]_{\geq 1}[S]_{\geq 1}[E/D]_{\leq 5}[S]_{\geq 1}[E/D]_{\geq 1}$ was used to search *S. cerevisiae* open reading frames listed in the Saccharomyces Genome Database, using the Yeast Genome Pattern Matching (PatMatch) algorithm (http://www.yeastgenome.org/cgi-bin/PATMATCH/nph-patmatch). This search yields 160 *S. cerevisiae* proteins that can possibly undergo $InsP_7$-mediated phosphorylation. This includes APL6, the *S. cerevisiae* homologue of the AP3 adaptor protein beta subunit AP3β3 (Fig. 2).

Once a protein sequence that can potentially be phosphorylated by $InsP_7$ is identified, the entire protein or a fragment encompassing the likely phosphorylation site (acidic serine sequence motif), needs to be overexpressed and purified for analysis. Two alternative strategies can be utilized for this: (a) express the protein in a eukaryotic host, such as *S. cerevisiae*, so that the protein gets prephosphorylated or (b) express the protein in *E. coli*, and subsequently prephosphorylate it in vitro using the protein kinase CK2, which phosphorylates serine residues in acidic serine motifs. Subheadings 3.2 and 3.3 describe these alternative strategies.

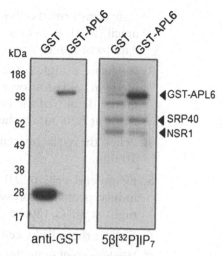

Fig. 2. InsP₇-phosphorylation of proteins expressed in *S. cerevisiae*. APL6 was overexpressed in *S. cerevisiae* as a GST-fusion protein. Extracts were prepared from yeast as described in Subheading 3.4.1, followed by phosphorylation by radiolabelled InsP₇ as outlined in Subheading 3.4.4. Proteins were resolved by SDS-PAGE and transferred to a PVDF membrane. InsP₇ phosphorylation of APL6 was detected by autoradiography (*right panel*), and the presence of the protein was confirmed by Western blotting with an anti-GST antibody (*left panel*). An extract from yeast expressing GST alone was used as a control. InsP₇-phosphorylated *S. cerevisiae* proteins present in the extract are detected in the autoradiogram, and include NSR1 and SRP40 (7).

3.2. Overexpression and Purification of a GST-Fusion Protein in Yeast

The candidate InsP₇ substrate that is expressed as a GST-fusion in yeast can be tested for InsP₇-mediated phosphorylation either directly using an extract prepared from yeast cells overexpressing the protein, or after purifying the protein on glutathione-agarose beads (see Note 3 to decide which may be your best option).

3.2.1. Preparation of Extract from Yeast Expressing a GST-Fusion Protein for Direct Use in an InsP₇ Phosphorylation Assay

Subclone the cDNA expressing either the entire protein of interest, or a fragment encompassing the potential pyrophosphorylation site into a plasmid that allows protein expression in a eukaryotic host. We recommend expressing the candidate substrate protein in *S. cerevisiae*, as a galactose-inducible GST fusion protein, using the plasmid pYESGEX6p2 (plasmid available upon request, see Note 4). For example, we subcloned the cDNA encoding APL6, a candidate InsP₇ substrate identified by sequence analysis of the *S. cerevisiae* proteome, into pYESGEX6p2 (Fig. 2).

1. Prepare competent *S. cerevisiae* cells using the Lithium acetate-SS carrier DNA–PEG method (10), or the Frozen-EZ yeast transformation kit from Zymo Research (see Note 5).

2. Transform the competent yeast with the expression plasmid and select transformants based on the auxotrophic marker carried on the plasmid. For example, when using the plasmid pYESGEX6p2, which carries the URA3 selection marker,

plate transformed cells on synthetic complete medium without uracil (SC-URA + Dex medium), and incubate at 30°C for 2 days to obtain colonies.

3. Inoculate a single colony into 10 mL of SC-URA + Dex liquid medium, and grow overnight at 30°C with shaking at 250 rpm, till the density of the culture, measured by monitoring absorbance at 600 nm reaches A_{600} 2–4.

4. Pellet the cells by centrifugation at $1,000 \times g$ at 25°C for 10 min.

5. Resuspend cells in 10 mL SC-URA + Gal at A_{600} 0.5–1, to induce protein expression, and incubate with shaking overnight at 30°C, till the culture has reached A_{600} 3–5.

6. Collect the cells by centrifugation at $1,000 \times g$ for 5 min.

7. Wash the cell pellet by resuspension in 10 mL water and centrifuge again at $1,000 \times g$ for 5 min.

8. Resuspend the pellet in 1 mL ice cold Lysis Buffer. Distribute the cell suspension into two microcentrifuge tubes and add 500 μL glass beads to each tube. Vortex the tubes for 3–5 min at 4°C to lyse the cells.

9. Centrifuge at $14,000 \times g$ at 4°C for 15 min to pellet the beads and cell debris. Collect and store the supernatant on ice. This is the extract to be used for the phosphorylation assay.

10. Quantify total protein using a standard protein estimation method (Bradford or BCA Method) and dilute the extract in Lysis Buffer, if required, to a final concentration of 2 mg/mL.

11. Proceed with the $InsP_7$ phosphorylation assay (Subheading 3.4.1), preferably on the same day.

3.2.2. Purification of a GST-Fusion Protein from Yeast

1–4. Follow steps 1–4 described in Subheading 3.2.1.

5. Resuspend the cells grown in SC-URA + Dex in 50–100 mL SC-URA + Gal at A_{600} 1, and incubate with shaking overnight at 30°C till the culture has reached A_{600} 3–5.

6. Centrifuge the culture at $1,000 \times g$ for 5 min. Wash the pellet in 10 mL chilled water and centrifuge again.

7. Resuspend cells in 5 mL chilled GST Purification Buffer A. Transfer the cell suspension to microcentrifuge tubes (0.75 mL per tube), and add 500 μL glass beads. Lyse the cells by vortexing for 3–5 min at 4°C.

8. Centrifuge the extract at $14,000 \times g$ for 15 min at 4°C.

9. Transfer the supernatant to a 15 mL conical tube, being careful to leave behind the glass beads. If glass beads are transferred along with the supernatant, centrifuge again at $1,000 \times g$ for 2 min and transfer the extract to a new 15 mL conical tube. Add Triton X-100 to obtain a final concentration of 1% (v/v).

10. See Subheading 2.2.2 for preparation of glutathione-agarose beads. See Note 2 on how to pipette agarose beads. Transfer 300 µL of glutathione-agarose beads: Buffer B (1:1 v/v) suspension, to the cell extract.

11. Keep the tube mixing by head over tail rotation at 4°C for 2 h.

12. Centrifuge the sample at 1,000×g for 3 min at 4°C to pellet the beads. Remove the supernatant and wash the beads with 1 mL GST Purification Buffer C. Transfer the sample to a microcentrifuge tube, mix by inverting the tube several times, and centrifuge at 14,000×g for 1 min at 4°C.

13. Repeat three more washes in 1 mL Buffer C, followed by four washes in 1 mL Buffer B. Finally wash two times with 1 mL PBS, resuspend the beads in an equal volume of PBS, and store at 4°C till ready for use in the phosphorylation assay. Purified protein on beads can be stored at 4°C for up to 1 week prior to the phosphorylation assay.

14. Run a 10–20 µL aliquot of the beads suspension on SDS-PAGE along with protein molecular weight markers, and stain the gel to check protein purity and to quantify the purified protein.

15. Proceed with the InsP_7 phosphorylation assay (Subheading 3.4.2).

3.3. Prephosphorylation of a Protein Expressed in E. coli to Prime it for InsP$_7$-Mediated Phosphorylation

If the candidate InsP_7 substrate protein is not easily expressed in *S. cerevisiae*, or if expression of the protein has already been standardized in *E. coli*, protein purified from *E. coli* can be "primed" for InsP_7-mediated phosphorylation by prephosphorylation with the protein kinases CK1 or CK2. CK1 has a sequence recognition motif of D/E/pSXXS and CK2 recognizes D/EXXS. Therefore, most serine residues in acidic serine sequence motifs can be phosphorylated by these protein kinases.

1. Overexpress the protein or protein fragment of interest as a fusion to GST, using a pGEX *E. coli* expression plasmid (GE Healthcare Life Sciences).

2. Purify the GST fusion protein using glutathione-agarose beads according to the protocol supplied by the manufacturer, and store the protein bound to beads in PBS (beads: PBS, 1:1 v/v) at 4°C.

3. Set up the prephosphorylation reaction with 10 µL protein-bound-glutathione-beads: PBS (1:1 v/v) suspension, 1 mM ATP, 1× kinase buffer supplied by the manufacturer, and 200–500 ng (approximately 250 units) purified protein kinase (CK2 holoenzyme, CK2 alpha subunit, or CK1) in a final volume of 25 µL.

4. Incubate the reaction at 30 or 37°C (as recommended by the manufacturer) for 1 h, mixing the beads intermittently.

5. Wash the beads with chilled PBS (two washes of 1 mL each), followed by two washes with 1 mL of 1× Phosphorylation Buffer.

6. Resuspend the beads in 1× Phosphorylation Buffer to a final volume of 25 μL, and store them on ice.

7. Proceed with the $InsP_7$ phosphorylation assay (Subheading 3.4.3).

3.4. Phosphorylation of Proteins by 5β[^{32}P]InsP$_7$

Subheadings 3.2 and 3.3 describe multiple ways to prepare a candidate $InsP_7$ substrate protein for phosphorylation. Phosphorylation of the protein of interest can be performed (a) directly using an extract from yeast cells overexpressing the protein (Subheading 3.2.1), (b) on GST-fusion protein purified from yeast and bound to glutathione-agarose beads (Subheading 3.2.2), or (c) on protein purified from *E.coli* that is prephosphorylated with CK1 or CK2 (Subheading 3.3). These alternative methods are described in Subheadings 3.4.1, 3.4.2, and 3.4.3, respectively.

5β[^{32}P]InsP_7 is prepared as described in Chapter 5, Subheading 3.1.3, to obtain a specific activity of 60 Ci/mmole. 10^6 CPM (approximately 1 μCi) 5β[^{32}P]InsP_7 is used per phosphorylation reaction (see Note 6).

3.4.1. InsP$_7$-Phosphorylation of a Cell Extract from Yeast Expressing a Candidate Protein

An extract obtained from yeast cells overexpressing a protein of interest can be used directly to determine whether the protein can be phosphorylated by $InsP_7$. See Fig. 2 for an example.

1. Set up a reaction using 12 μg (6 μL of 2 mg/mL yeast cell extract, obtained in Subheading 3.2.1), 10^6 CPM 5β[^{32}P]InsP_7, and 2 μL of 10× Phosphorylation Buffer in a final volume of 20 μL.

2. Incubate the reaction at 37°C for 15 min.

3. Add 7 μL of 4× NuPAGE sample buffer, mix, and heat the sample at 95°C for 5 min.

4. Run the samples on a NuPAGE gel along with prestained protein molecular weight markers. Go to Subheading 3.4.4.

3.4.2. Phosphorylation of a GST-Fusion Protein Expressed and Purified from S. cerevisiae

1. Mix 10–15 μL of purified GST-fusion protein bound to glutathione-agarose beads in a PBS suspension (obtained in Subheading 3.2.2), 10^6 CPM 5β[^{32}P]InsP_7 in a final volume of 20–25 μL of 1× Phosphorylation Buffer.

2. Incubate the reaction at 37°C for 15 min.

3. Add 7 μL of 4× NuPAGE sample buffer, mix, and heat the sample at 95°C for 5 min.

4. Run the samples on a NuPAGE gel. Go to Subheading 3.4.4.

3.4.3. Phosphorylation of a "Primed" GST-Fusion Protein Purified from E. coli

The example shown in Fig. 3 describes InsP$_7$ phosphorylation of a fragment of *S. cerevisiae* APL6 after it has been prephosphorylated by protein kinases CK1 or CK2.

1. To the prephosphorylated GST-fusion protein purified from *E. coli* (obtained in Subheading 3.3), which is resuspended in 25 μL of 1× Phosphorylation Buffer, add 10^6 CPM 5β[^{32}P] InsP$_7$.

2. Incubate the reaction at 37°C for 15 min.

3. Add 8.5 μL of 4× NuPAGE sample buffer, mix, and heat the sample at 95°C for 5 min.

4. Run the samples on a NuPAGE gel. Go to Subheading 3.4.4.

3.4.4. Detection of InsP$_7$-Phosphorylated Protein by Western Blotting and Autoradiography

1. Run the gel until radiolabelled InsP$_7$ comes out of the gel. This can be ensured by running the gel until the 3 kDa prestained marker band has run out of the bottom of the gel.

2. Transfer the proteins to a PVDF membrane by Western blotting.

3. Dry and expose the membrane to an X-ray film or phosphorimager plate to detect the radiolabelled proteins by autoradiography (see Note 7).

4. Rehydrate the membrane and proceed with immunoblotting using an anti-GST antibody. Align the autoradiogram and immunoblot to determine whether your protein of interest has been phosphorylated by InsP$_7$ (Fig. 2) (see Note 8 and Fig. 3).

3.5. Testing the Biochemical Properties of InsP$_7$-Phosphorylated Proteins

To distinguish pyrophosphoserine, generated upon InsP$_7$-mediated phosphorylation, from phosphoserine generated by protein kinase-mediated phosphorylation, we recommend testing the InsP$_7$-phosphorylated protein for sensitivity to acid and phosphatases. The β phosphate moiety of pyrophosphoserine is cleaved in the presence of 3 N HCl at 37°C, whereas phosphoserine is stable under these conditions (8). In contrast, phosphoserine is converted to serine by treatment with lambda protein phosphatase, whereas pyrophosphoserine is inert to this treatment.

This section describes methods to test these biochemical properties of pyrophosphoproteins generated by InsP$_7$-mediated phosphorylation. The protein of interest phosphorylated by a protein kinase (usually CK2) is required as a control for these assays. These tests are best performed with protein that has been overexpressed and purified from *E. coli*, so that there is minimal interference from other posttranslational modifications that occur in eukaryotic cells.

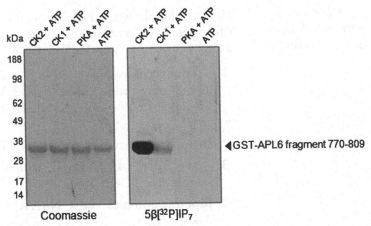

APL6 770-809:   VEESSDEDEDESEESSDDDEYSDSSLGTSSSGTSSSHLEL

Fig. 3. InsP$_7$-phosphorylation of proteins purified from *E. coli*. The acidic serine motif present in *S. cerevisiae* APL6 (amino acid residues 770-809) was expressed as a GST-fusion protein in *E. coli* and purified on glutathione-agarose beads. Purified protein was preincubated with the indicated protein kinases and ATP (Subheading 3.3), the beads were washed with PBS, and then incubated with radiolabelled InsP$_7$ (Subheadings 3.4.3 and 3.4.4). Proteins were resolved by SDS-PAGE, the gel was stained (*left panel*), dried and radiolabelled proteins were detected by autoradiography (*right panel*). Protein kinases CK2, and to a lower extent, CK1 are able to pre-phosphorylate APL6 aa 770-809, whereas Protein Kinase A (PKA) does not.

Simultaneously set up reactions according to Subheadings 3.5.1 and 3.5.2, to generate radiolabelled phosphorylated and pyrophosphorylated protein.

3.5.1. CK2-Mediated Phosphorylation of the Candidate Protein Expressed in E. coli

1. Start with your protein of interest expressed in *E. coli* as a GST-fusion and purified as described in Subheading 3.3, steps 1 and 2.

2. Set up the protein kinase reaction with 40 μL of protein-bound-glutathione beads: PBS (1:1 v/v) suspension, 200 μM Mg-ATP, 4 μCi γ[^{32}P]ATP, 1–2 μg (approximately 1,000 units) purified protein kinase (CK2 holoenzyme, CK2 alpha subunit, or CK1), 10 μL of 10× kinase buffer supplied by the manufacturer, in a final volume of 100 μL (see Note 6).

3. Incubate at 30 or 37°C, as recommended by the manufacturer, for 15 min.

4. Stop the reaction by adding EDTA to a final concentration of 5 mM.

5. Add 1 mL chilled PBS, mix thoroughly, and centrifuge at 14,000×*g* for 2 min at 4°C to pellet the glutathione-agarose beads. Remove the supernatant carefully, making sure not to aspirate any beads. Discard the radioactive supernatant appropriately (see Note 6).

6. Repeat two more washes with chilled PBS.

7. Resuspend the beads (approximately 20 µL bed volume) in an equal volume of PBS.

3.5.2. InsP₇-Mediated Phosphorylation of the Candidate Protein Expressed in E. coli

1. Start with your protein of interest expressed in *E. coli* as a GST-fusion protein, purified and prephosphorylated as described in Subheading 3.3, steps 1–6.

2. Set up the InsP₇ phosphorylation reaction with 40 µL protein-bound-glutathione beads: PBS (1:1 v/v) suspension, 4 µCi (4×10^6 CPM) $5\beta[^{32}P]$InsP₇, 10 µL of 10× Phosphorylation Buffer, in a final volume of 100 µL (see Note 6).

3. Incubate the reaction at 37°C for 15 min, followed by 95°C for 5 min (see Note 9).

4. Add EDTA to a final concentration of 5 mM to terminate the reaction.

5. Wash and prepare the beads as described in steps 5–7 in Subheading 3.5.1.

3.5.3. Testing Acid Lability of InsP₇-Phosphorylated Protein

The two sets of radiolabelled protein prepared as described above are (a) CK2 + γ[^{32}P]ATP phosphorylated GST-fusion protein on agarose beads: PBS (1:1 v/v), Subheading 3.5.1, and (b) $5\beta[^{32}P]$InsP₇-phosphorylated GST-fusion protein on agarose beads: PBS (1:1 v/v), Subheading 3.5.2.

1. Prepare two aliquots, "control" and "acid treatment" for each sample (a) and (b), by adding 10 µL of the agarose beads: PBS suspension per tube.

2. In each set, to the "control" tube, add 10 µL water, and to the "acid treatment" tube, add 10 µL of 6 N HCl (to obtain a final concentration 3 N HCl). Incubate the samples at 37°C for 1 h.

3. To the "control" tube, add 6 µL water. To the "acid treatment" tube, add 6 µL of 10 M NaOH to neutralize the sample (see Note 10).

4. Add 9 µL of 4× LDS buffer to all samples. Incubate at room temperature for 5 min. Do not heat the sample.

5. Run the samples on NuPAGE, and stain the gel to detect purified proteins (see Note 8).

6. Dry the gel and determine the phosphorylation status of proteins by autoradiography. The phosphates added by InsP₇ will be removed upon treatment with acid, whereas phosphate added by CK2 + ATP will be resistant to acid treatment.

3.5.4. Testing Phosphatase Resistance of InsP₇-Phosphorylated Protein

The two samples to be tested are (a) CK2 + γ[^{32}P]ATP phosphoylated GST-fusion protein on agarose beads: PBS (1:1 v/v), Subheading 3.5.1, and (b) $5\beta[^{32}P]$InsP₇-phosphorylated GST-fusion protein on agarose beads: PBS (1:1 v/v), Subheading 3.5.2.

1. Prepare two aliquots, "control" and "phosphatase treatment" for each sample (a) and (b), by adding 10 μL of the agarose beads: PBS suspension per tube.

2. To the "control" sample, add 2.5 μL of 10× lambda phosphatase buffer and water to a final volume of 25 μL. To the "phosphatase treatment" sample, add 2.5 μL of 10× lambda phosphatase buffer, 400 units of lambda protein phosphatase, and water to a final volume of 25 μL.

3. Incubate the samples at 37°C for 1 h.

4. Add 9 μL of 4× LDS buffer to all samples. Incubate at room temperature for 5 min. Do not heat the sample.

5. Run the samples on NuPAGE, and stain the gel to detect purified proteins (see Note 8).

6. Dry the gel and determine the phosphorylation status of proteins by autoradiography. The phosphate added by $InsP_7$ will be resistant, whereas the phosphate transferred from ATP by CK2 will be removed upon treatment with lambda protein phosphatase.

4. Notes

1. All buffers and media should be prepared with ultra-pure deionized water (resistivity of 18.2 megohm/cm) and stored at 4°C, unless otherwise stated.

2. To accurately pipette a suspension of glutathione-agarose beads, always cut the end of a 200 μL or 1 mL pipette tip before use.

3. If your protein of interest does not comigrate with the major $InsP_7$-phosphorylated *S. cerevisiae* proteins, i.e., NSR1 (60 kDa) and SRP40 (63 kDa), it is possible to test $InsP_7$-mediated phosphorylation without purifying the protein, by directly using the extract obtained from yeast overexpressing the protein.

4. The yeast expression plasmid pYESGEX6p2 has been constructed using the pYES2NTA (Invitrogen) backbone, by inserting the GST fusion expression cassette from pGEX6p2 (GE Healthcare Life Sciences).

5. Although any strain of *S. cerevisiae* can be used for the protocol described, we prefer to use the protease deficient strain DDY1810, (*MAT*a *leu2 ura3-52 prb1-112 pep4-3 pre1-451*).

6. Take appropriate precautions when handling radioactivity, and always follow the radiation safety guidelines and good laboratory practice rules laid down by your laboratory.

7. Be careful to mark the positions of the prestained markers on the autoradiogram to ensure proper alignment of the autoradiogram bands with the bands obtained after immonoblotting. This can be done by using radioactive ink or a luminescent marker such as Glogos® II Autorad Markers (Stratagene, 420201).

8. As an alternative to Western blotting, if a purified GST-fusion protein is used for the InsP$_7$ phosphorylation assay, the gel can be stained with Coomassie Blue stain, dried, and subjected to autoradiography. This will work especially well if large amounts of pure GST-fusion protein are obtained by overexpression in *E. coli*. Stain the gel by immersing it in Coomassie Blue staining solution for 30 min, while shaking at room temperature. Remove excess stain by rinsing in water and incubate the gel in destaining solution until the bands are clear. Since the gel is radioactive, also see Note 6.

9. Being an enzyme-independent chemical reaction, in vitro, the extent of InsP$_7$-mediated protein phosphorylation increases with increasing temperature (8). Although phosphorylation is observed at 37°C, an additional incubation step at 95°C maximizes the yield of pyrophosphorylated protein required for subsequent analysis. We suggest carrying out a time course of phosphorylation of your purified substrate protein by 5β[^{32}P]InsP$_7$ at 37°C. Note that during such an assay, after incubation with 5β[^{32}P]InsP$_7$, the reaction should be mixed with SDS sample buffer at room temperature, and not be heated prior to SDS-PAGE.

10. Carry out trials to determine the volume of 10 M NaOH required to neutralize 20 µL of 3 M HCl.

Acknowledgements

We thank Adolfo Saiardi and Solomon H. Snyder for suggestions and helpful comments. This work is supported by the Department of Biotechnology, Government of India, and The Wellcome Trust, UK.

References

1. Bennett, M., Onnebo, S. M., Azevedo, C., and Saiardi, A. (2006) Inositol pyrophosphates: metabolism and signaling. *Cell Mol Life Sci* **63**, 552–564.

2. Burton, A., Hu, X., and Saiardi, A. (2009) Are inositol pyrophosphates signalling molecules? *J Cell Physiol* **220**, 8–15.

3. Illies, C., Gromada, J., Fiume, R., Leibiger, B., Yu, J., et al. (2007) Requirement of inositol pyrophosphates for full exocytotic capacity in pancreatic cells. *Science* **318**, 1299–1302.

4. Bhandari, R., Juluri, K. R., Resnick, A. C., and Snyder, S. H. (2008) Gene deletion of inositol hexakisphosphate kinase 1 reveals

inositol pyrophosphate regulation of insulin secretion, growth, and spermiogenesis. *Proc Natl Acad Sci USA* **105,** 2349–2353.

5. Luo, H. R., Huang, Y. E., Chen, J. C., Saiardi, A., Iijima, M., et al. (2003) Inositol pyrophosphates mediate chemotaxis in Dictyostelium via pleckstrin homology domain-PtdIns(3,4,5)P3 interactions. *Cell* **114,** 559–572.

6. Lee, Y. S., Huang, K., Quiocho, F. A., and O'Shea, E. K. (2008) Molecular basis of cyclin-CDK-CKI regulation by reversible binding of an inositol pyrophosphate. *Nat Chem Biol* **4,** 25–32.

7. Saiardi, A., Bhandari, R., Resnick, A. C., Snowman, A. M., and Snyder, S. H. (2004) Phosphorylation of proteins by inositol pyrophosphates. *Science* **306,** 2101–2105.

8. Bhandari, R., Saiardi, A., Ahmadibeni, Y., Snowman, A. M., Resnick, A. C., et al. (2007) Protein pyrophosphorylation by inositol pyrophosphates is a posttranslational event. *Proc Natl Acad Sci USA* **104,** 15305–15310.

9. Yan, T., Yoo, D., Berardini, T. Z., Mueller, L. A., Weems, D. C., et al. (2005) PatMatch: a program for finding patterns in peptide and nucleotide sequences. *Nucleic Acids Res* **33,** W262–W266.

10. Gietz, R. D., and Woods, R. A. (2002) Transformation of yeast by lithium acetate/single-stranded carrier DNA/polyethylene glycol method. *Meth Enzymol* **350,** 87–96.

Chapter 7

Synthesis and Nonradioactive Micro-analysis of Diphosphoinositol Phosphates by HPLC with Postcolumn Complexometry

Hongying Lin, Karsten Lindner, and Georg W. Mayr

Abstract

A nonradioactive high-performance anion-exchange chromatographic method based on MDD-HPLC (Mayr Biochem. J. 254:585–591, 1988) was developed for the separation of inositol hexakisphosphate ($InsP_6$, phytic acid) and most isomers of pyrophosphorylated inositol phosphates, such as diphospho-inositol pentakisphosphate ($PPInsP_5$ or $InsP_7$) and bis-diphosphoinositol tetrakisphosphate (bis$PPInsP_4$ or $InsP_8$). With an acidic elution, the anion-exchange separation led to the resolution of four separable $PPInsP_5$ isomers (including pairs of enantiomers) into three peaks and of nine separable bis$PPInsP_4$ isomers into nine peaks. The whole separation procedure was completed within 20–36 min after optimization. Reference standards of all bis$PPInsP_4$ isomers were generated by a nonenzymatic shotgun synthesis from $InsP_6$. Hereby, the phosphorylation was brought about nonenzymatically when concentrated $InsP_6$ bound to the solid surface of anion-exchange beads was incubated with creatine phosphate under optimal pH conditions. From the mixture of pyrophosphorylated $InsP_6$ derivatives containing all theoretically possible isomers of $PPInsP_5$, bis$PPInsP_4$, and also some isomers of tris$PPInsP_3$, isomers were separated by anion-exchange chromatography and fractions served as reference standards of bis$PPInsP_4$ isomers for further investigation. Their isomeric nature could be partly assigned by comparison with position specifically synthesized or NMR-characterized purified protozoan reference compounds and partly by limited hydrolysis to $PPInsP_5$ isomers. By applying this nonradioactive analysis technique to cellular studies, the isomeric nature of the major bis$PPInsP_4$ in mammalian cells could be identified without the need to obtain sufficient material for NMR analysis.

Key words: Inositol phosphate, Pyrophosphates, Signal transduction, Metal-dye detection, HPLC

1. Introduction

Myo-inositol hexakisphosphate ($InsP_6$, Fig. 1a) is the most abundant inositol phosphate in eukaryotic cells and tissues, with a concentration of 10–60 μM in mammalian cells (2), a range

Christopher J. Barker (ed.), *Inositol Phosphates and Lipids: Methods and Protocols*, Methods in Molecular Biology, vol. 645,
DOI 10.1007/978-1-60327-175-2_7, © Humana press, a part of Springer Science+Business Media, LLC 2010

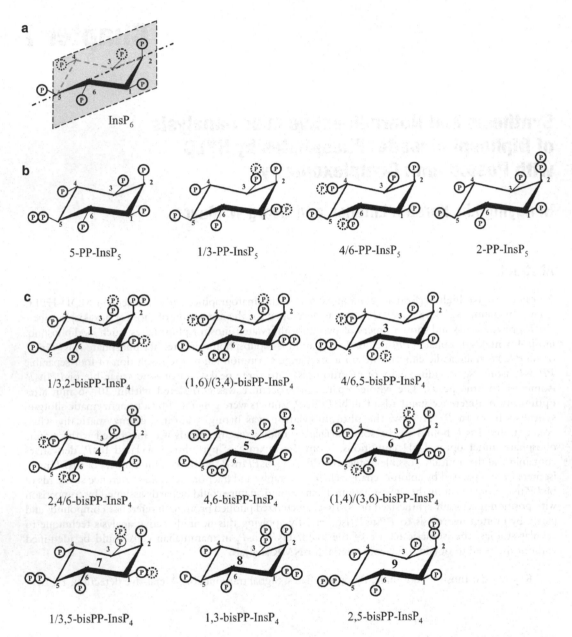

Fig. 1. Structures of InsP$_6$ and pyrophosphorylated inositol phosphates. (a) myo-inositol hexakisphosphate (InsP$_6$). Inositol phosphate nomenclature recognizes both the number of attached phosphate ("P") or pyrophosphate groups, and their positions around the inositol ring. Only the 2-phosphate is axial towards the plane of the inositol ring; the other five phosphates are equatorial (see (a)). Therefore, two pairs of enantiomers exist for PPInsP$_5$ isomers and six enantiomeric pairs for bisPPInsP$_4$. (b) All four HPLC separable diphosphorylated inositol phosphates (PPInsP$_5$ or InsP$_7$). (c) All nine HPLC separable bis-diphosphorylated inositol phosphates (bisPPInsP$_4$ or InsP$_8$). Dotted rings around P in (b) and (c) denote positions of pyrophosphates giving rise to an enantiomeric pair, which cannot be separated by anion-exchange HPLC. In each case, these pairs can be mirror imaged by a plane extending vertically from the ring level through phosphate groups 2 and 5, respectively (see Fig. 1a).

between 10 and 15 nmol/g wet weight in distinct rat brain regions (3) and about 0.7 mM in slime molds (4). $InsP_6$ is the predominant precursor of inositol pyrophosphates such as diphosphoinositol pentakisphosphate ($PPInsP_5$ or $InsP_7$, Fig. 1b) and bis-diphosphoinositol tetrakisphosphate (bis$PPInsP_4$ or $InsP_8$, Fig. 1c) (5). In addition, $InsP_5$ and $InsP_4$ isomers have also been identified as precursors of bis$PPInsP_3$ and bis$PPInsP_4$ (6). Diphosphoinositol phosphates were first discovered in the *Dictyostelium discoideum* wild-type strain NC4 (7) and have been structurally resolved since 1992 (8–11). The predominant isomer of $PPInsP_5$ detected in *Polysphondylium* (12), *Entamoeba histolytica* (13), and in several mammalian cell types (10) has the diphosphate group attached to the 5-position (5-$PPInsP_5$), whereas the major isomer of $PPInsP_5$ in *Dictyostelium* was identified as D-6-$PPInsP_5$ (11). Two isomers of bis$PPInsP_4$ have been identified by 2D NMR analysis, namely 5,6-bis$PPInsP_4$ in *Dictyostelium* (11) and *Polysphondylium* species (12), and an additional isomer, 1,5-bis$PPInsP_4$, or its corresponding enantiomer, 3,5-bis$PPInsP_4$, in *Polysphondylium* (12).

The cellular functions of inositol pyrophosphates and their metabolism pathway in different organisms were reviewed by Bennett and co-workers (14), York (15), and Irvine (16). The important functions described include DNA hyper-recombination (17), chemotaxis in *Dictyostelium* (18), telomere maintenance (19, 20), mRNA export (21), exocytosis (22), and protein phosphorylation (23). Choi and co-workers reported that bis$PPInsP_4$ is a novel sensor of mild thermal stress in mammalian cells (24). Furthermore, in mammalian cells, bis$PPInsP_4$ accumulates rapidly in response to hyperosmotic stress (25). Recently, it was shown that certain isomers of $InsP_6$ kinase at least in vitro are able to generate even triphosphate groups at certain positions of the inositol ring (26).

In our laboratory, a high yield nonenzymatic shotgun synthesis of $PPInsP_5$, bis$PPInsP_4$, and tris$PPInsP_3$ isomers from $InsP_6$ was developed. Nonenzymatic pyrophosphorylation occurs already when solutions of inositol hexakisphosphate are freeze-dried in the presence of a high-energy phosphate, such as creatine phosphate (CrP), under acidic condition (8). However, the yield never exceeds 5%. A dramatically higher yield up to 50% could be obtained when concentrated $InsP_6$ bound to the surface of tightly packed Q-Sepharose beads in a column was directly attacked by concentrated CrP under acidic pH conditions. The mixture was purified on the same Q-Sepharose column by acidic elution and fractions served as standards of bis$PPInsP_4$ for further investigation. The goals of this study were: (1) to develop a method where diphosphoinositol phosphate isomers are nonenzymatically synthesized from $InsP_6$; (2) to establish a reliable nonradioactive analytical method that can separate nanomole amounts of $InsP_6$

and most of its pyrophosphorylated derivates, containing up to nine phosphates per inositol; (3) to clearly identify the three biologically relevant out of the nine theoretically separable isomers of bisPPInsP_4 by MDD-HPLC based comparison with isomers specifically synthesized or obtained from previously isolated, NMR-characterized, biological reference compounds (with or without partial hydrolysis); and (4) to prepare a reproducible in-house reference standard mixture suitable to identify further particular bis-diphosphoinositol phosphate and tris-diphosphoinositol phosphate isomers containing up to nine phosphates per inositol ring and even more complex isomers that may be discovered in mammalian cells.

2. Materials

Water for preparation of all solutions was purified from predeionized water by a Millipore Milli-Q system. All chemicals were of analytical grade or higher purity unless otherwise stated.

2.1. Nonenzymatic Synthesis and Purification of Diphosphoinositol Phosphates

1. Phytic acid (dodecasodium salt hydrate; Aldrich, Milwaukee, WI, USA).
2. Creatine phosphate (disodium tetrahydrate; ICN Biomedicals Inc., Ohio, USA).
3. A speed vacuum centrifuge (Bachhofer, Reutlingen, Germany).
4. Q-Sepharose (Fast Flow, 300 mL, Code No. 17-0510-01), or a cross-linked Q-Sepharose column (XK16/20) (Pharmacia Biotech, Uppsala, Sweden). The bead matrix active groups on both materials are $-O-CH_2CHOHCH_2OCH_2CHOHCH_2N^+(CH_3)_3$ and the particle pore size is 34 μm (Pharmacia Biotech, Uppsala, Sweden).
5. An ÄKTAprime FPLC control system (Pharmacia Biotech, Uppsala, Sweden).

2.2. Determination of Total Inorganic Phosphate

1. Malachite green hydrochloride (Sigma, MO, USA).
2. Tergitol NP-10 (Sigma, MO, USA).
3. Ammonium heptamolybdate (Sigma, MO, USA).
4. 0.045% (w/v) malachite green oxalate aqueous solution: Dissolve 0.45 g of malachite green oxalate in 1 mL of deionized water. Store in a dark brown bottle.
5. 4.2% ammonium heptamolybdate in 4 M HCl solution: Dissolve 4.2 g of ammonium heptamolybdate in 100 mL of 4 M HCl. Store in a plastic bottle away from direct sunlight.

6. Stock Pi standard solution (20 mM): 0.136-g KH_2PO_4 is dissolved in 50 mL 1 N H_2SO_4 solution.

7. A 250-mL working reagent solution consists of: 183.75-mL malachite green (0.045% w/v) and 61.25-mL ammonium heptamolybdate (4.2% w/v, in 4 M HCl). The solution is mixed for 24 h, then 5-mL Tergitol NP-10 (2% v/v) is added.

2.3. HPLC Analysis

1. Analytical grade HCl (30%, suprapure; Merck, Darmstadt, Germany).

2. Triethanolamine (TEA, p.a., >99% purity; Merck, Darmstadt, Germany).

3. 4-(2-pyridylazo)-resorcinol monosodium salt monohydrate (PAR; Merck, Darmstadt, Germany).

4. Methanol (LiChrosolv; Merck, Darmstadt, Germany).

5. Yttrium trichloride hexahydrate (Aldrich, Milwaukee, WI, USA).

6. Sodium fluoride (suprapure; Sigma, MO, USA).

7. Sodium acetate (Merck, Darmstadt, Germany).

8. A 10-mM stock solution of PAR: Dissolve 2.55 g of PAR in 1 L methanol and stored in a plastic bottle at –20°C.

9. A 18-mM stock solution of YCl_3: Dissolve 1.092 g of YCl_3 in 200 mL of deionized water. Store in a dark brown bottle at 4°C.

10. A 0.5-M stock solution of sodium acetate: Dissolve 4.10 g of sodium acetate in 100 mL of deionized water and store at 4°C.

11. A 0.5-M stock solution of sodium fluoride (NaF): Dissolve 2.10 g of NaF in 100 mL of deionized water and store at 4°C.

12. MDD-HPLC eluent A (0.2 mM HCl, 15 µM YCl_3): 21 µL 30% HCl, 833 µL 18 mM YCl_3, filled with H_2O to 1 L.

13. MDD-HPLC eluent B (0.5 M HCl, 15 µM YCl_3): 52.9 mL, 30% HCl, 833 µL 18 mM YCl_3, filled with H_2O to 1 L.

14. Postcolumn reagent C (1.6 M triethanolamine, 300 µM PAR, pH 9.0): Dissolve 238.7 g triethanolamine in 900 mL of deionized water, add 30 µL 10 mM PAR, adjust pH with 30% HCl (~11 mL), then fill up with H_2O to 1 L.

15. Prior to use, solvents and the postcolumn reagent are filtered and degassed by vacuum filtration through inert 0.22-µm pore size membrane filters (Millipore, type GV) in a Pyrex glass filtration device.

16. HPLC injection solution (2 mM sodium acetate, 2 mM NaF): 4 mL sodium acetate (0.5 M) and 4 mL NaF (0.5 M) filled with H_2O to 1 L and filtered. Store at 4°C.

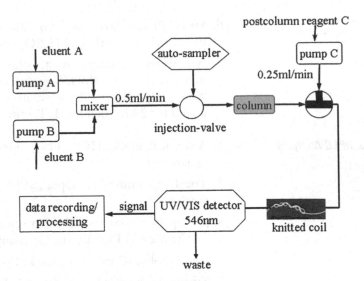

Fig. 2. Scheme of a HPLC system for metal-dye detection.

17. The basic setup of an automatized HPLC system suitable for an inositol phosphate micro-analysis with a postcolumn metal-dye detection (MDD) is depicted in Fig. 2.

18. This HPLC system consists of two pumps (Pump 422, Kontron) for gradient elution, a further pump for postcolumn dye reagent (LC-10 AD, Shimadzu), a HPLC auto-sampler 560 (Kontron) with an injection loop volume of 1 mL and a loading syringe of 2.5 mL (Hamilton, Swiss), an UV/VIS recorder (SPD-10Avvp, Shimadzu) for absorbance recording, and a Galaxie Chromatography Data System (Varian, Palo Alto, CA, USA) for control and data processing. For delivery of both solvent and postcolumn reagent, true double-piston pumps with (identical) stroke volumes below 80 µL are essential. All eluent-wetted parts in this or analogous systems have to be inert (i.e., from titanium, sapphire, PEEK, and Teflon). In addition, all fittings, filter units, column inlets, and column frits are made from inert polymer such as PEEK or Teflon.

19. Mini Q PC 3.2/3 column (3 µm bead diameter) is purchased from Pharmacia Biotech (Uppsala, Sweden).

20. A knitted coil is handmade by a 40 cm 1/16″×0.5 mm ID PTFE capillary with seven knots (CS-Chromatographie Service GmbH, Langerwehe, Germany).

2.4. Extraction of Cellular Inositol Phosphates and Diphosphoinositol Phosphates

1. Dulbecco's Modified Eagle Media (DMEM; Invitrogen, Karlsruhe) is supplemented with 10% fetal bovine serum (FBS; Invitrogen, Karlsruhe).

2. Trichloroacetic acid (crystal extra pure; Merck, Darmstadt, Germany).

3. Diethyl ether (Merck, Darmstadt, Germany).

4. 8% (w/v) trichloroacetic acid solution: Dissolve 16 g of trichloroacetic acid in 200 mL of deionized water. Store at 4°C.

5. Water-saturated diethyl ether: Prepared by vigorously mixing 1 volume deionized water with 2 volume diethyl ether for at least 2 min.

6. 1 M triethanolamine: Dissolve 7.45 g TEA in 50 mL of deionized water. Store at 4°C.

7. Cell scrapers (Merck, Darmstadt, Germany).

3. Methods

3.1. Nonenzymatic Shotgun Synthesis and Purification of Diphosphoinositol Phosphates

Isomers of diphosphoinositol phosphates are shotgun synthesized in batch by a nonenzymatic, dehydration driven transphosphorylation reaction in which the high-energy phosphate of CrP is transferred to $InsP_6$ at pH $\leq$ 3 similar as described (8) but with essential improvements increasing the yield from $\leq$ 5 to 50%.

1. Transfer 100–500 µmol $InsP_6$ (dodecasodium hydrate, dissolved in water), 1 mmol CrP, and 2 mmol NaCl into a centrifuge tube with screw cap, then add water for a final volume of 25 mL, and adjust pH ca. 3 with HCl.

2. The tube is tilted for 5 min before the mixture is frozen in liquid nitrogen and subsequently dehydrated by overnight freeze-drying. Nonenzymatic phosphorylation of inositol phosphates occurs during dehydration, and the possible reactions are as shown in Fig. 3a.

3. Freeze-dried material, containing creatine, CrP, $InsP_6$, $PPInsP_5$, and bis$PPInsP_4$, is diluted with water (about 50–200 mL) to a conductivity of $\leq$ 3 mS.

4. Load onto an anion-exchange column (16×200 mm packed with Q-Sepharose fast flow) with a flow rate of 1 mL/min (see Note 1).

5. Wash with 20 mL of 0.2 M HCl at the same flow rate.

6. Gradient elute at the same flow rate with HCl rising in concentration from 0.2 to 0.5 M within 2 h.

7. The column is finally washed with 20 mL 0.5 M HCl.

8. Fractions of 0.5 mL are collected throughout.

9. Precisely pipette 50 µL of sample from each fraction of 0.5 mL into a glass tube for determination of total phosphate concentration according the malachite green/phosphomolybdate based method (see Subheading 3.2 and Fig. 4).

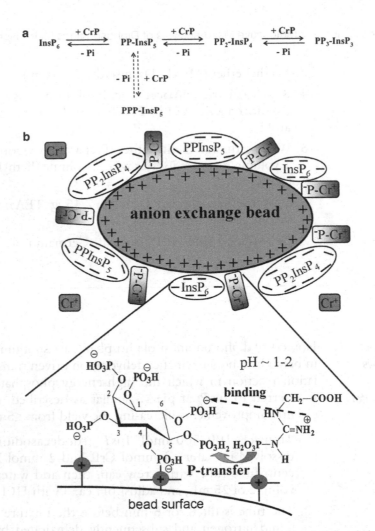

Fig. 3. Hypothetical mechanism of high yield solid anion-exchange-phase based diphosphoinositol phosphates synthesis from $InsP_6$ in the presence of zwitterionic high-energy phosphate donor, creatine phosphate. (a) Possible reactions of $InsP_6$ and creatine phosphate to diphosphoinositol phosphates and PPP $InsP_5$; (b) Possible mechanism for high yield transphosphorylation from creatine phosphate to inositol phosphates on the surface of polycationic beads in an anion-exchange column under suitable acidic washing conditions.

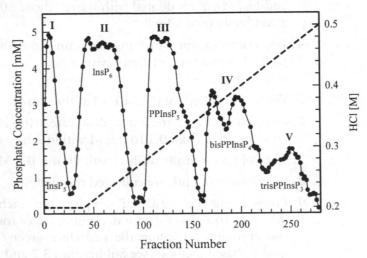

Fig. 4. Separation of diphosphoinositol phosphates synthesized on Q-Sepharose beads by chromatography on the same beads. *Dots* show the total phosphate concentration in each fraction, and the *dashed line* depicts concentration of eluent HCl.

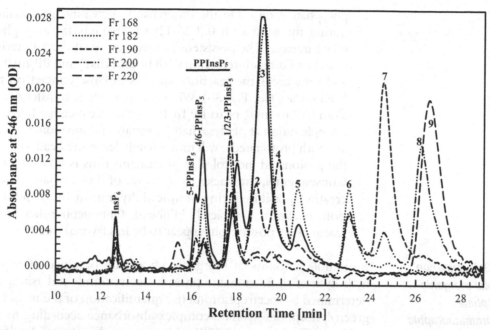

Fig. 5. MDD-HPLC analysis of several Q-Sepharose fractions containing diphosphoinositol phosphates (Gradient II). The nine HPLC separable bisPPInsP_4 isomers are marked by numbers 1–9.

10. After there, fractions are neutralized with 1 M triethanolamine base and stored at –20°C.

11. Isomerism of PP$_x$Ins$P_{(6-x)}$ in each fraction is determined by MDD-HPLC analysis (2–30 µL of sample from each fraction, see Subheading 3.3 and Fig. 5).

12. Fractions of individual isomers are pooled and stored at –20°C for further investigation (e.g., as reference standard, re-chromatograph, and partial hydrolysis).

13. Note: The dramatic increase in yield of pyrophosphorylated products from less than 5 % to about 50 % when performing the chromatography on Q-Sepharose is likely due to a solid-phase synthetic system, thereby established tightly packed on beads of the column head (Fig. 3b, see Note 2). In this system, InsP_6, PPInsP_5, and PP$_x$Ins$P_{(6-x)}$ are all concentrated and tightly bound to the quaternary ammonium groups of the beads until gradient elution is performed. This is also the case when the column is washed with 0.2 M HCl (pH about 0.7 (27)) due to the low pK values for the second protonation step of phosphate esters and pyrophosphate β-phosphates (28). Part of these phosphates becomes fully protonated under these conditions, i.e., suited for pyrophosphate bond formation. The zwitterion CrP has about 1.5 negative and 1 positive charge at pH 4 (pKs of acetate) and is still weakly bound to the beads at between pH 3 and about 1 due to its still partially unprotonated

phosphate group. On the other hand, increasing protonation during the wash with 0.2 M HCl is strongly favoring phosphate transfer. The persistent positive group of the zwitterion does favor direct interaction with bound $InsP_6$ and $PP_xInsP_{(6-x)}$ and thus keeps the reaction partners in close contact on the bead surface (see Fig. 3b). Whenever phosphate transfer occurs from CrP to $InsP_6$ or to $PP_xInsP_{(6-x)}$, the free creatine becomes a simple cation at pH less than 3 (acetate and guanidine groups are both protonated), will immediately leave the bead and also the protonated inositol phosphates and thus is washed away. Consequently, the backward reaction of β-phosphate back to creatine, possible during simple dehydration by lyophilization, is almost completely inhibited. Only dephosphorylation reactions are possible but appear to be less favored (see Note 2).

3.2. Determination of Total Inorganic Phosphate in Chromatographic Fractions

Nanomole amounts of inorganic phosphate obtained after acid hydrolysis of all fractionated phospho-compounds and ashing are determined by spectrophotometric quantification of the malachite green/phosphomolybdate complex absorbance according to the method of Lanzetta et al. (29) with some modifications. A sulfuric acid digestion step is employed for destruction of organic phosphate compounds and ashed organic residues are oxidized and removed by an H_2O_2 treatment. The test volume is reduced to a microplate scale (150 μL in our case) to handle a great number of probes simultaneously and to increase detection sensitivity. The procedure is as follows:

1. Precisely pipette 50 μL of sample from each fraction of 0.5 mL into a glass tube (12 × 100; Schott Duran, Germany).
2. Pipette 100 μL of 10 N H_2SO_4 (highest purity), mix well, and incubate for 1 h at 170°C in an oven.
3. Add 50 μL of 30% H_2O_2 (highest purity, Pi free), and incubate for another 20 min at 170°C till total evaporation of H_2O_2.
4. Dilute sample with water, and precisely adjust to a volume of 1 mL.
5. Mix 50 μL of diluted sample or standard solution with 100 μL working reagent.
6. Add 20 μL of 34% (w/v) trisodium citrate after 10 min to stop the reaction.
7. Wait for at least 20 min to allow the color to develop completely, and read the absorbance at 595 nm with a spectrophotometer.
8. The phosphate standard curve is linear over a wide concentration range, from ~0.5 to 10 nmol Pi in our case.
9. Volumes in steps 1, 4, 5, and 6 have to be adjusted or pipetted within an error rate of 1% to keep the total error within ±5%.

Multiple assays ($n \geq 3$) from each fraction with varied sample volumes are recommended.

10. For each fraction of 0.5 mL, the total phosphate is determined and five major peaks are detected (Fig. 4).

11. By detailed MDD-HPLC analysis of individual fractions and comparison with $InsP_5$ and $InsP_6$ standards and mixtures containing isomers of $PPInsP_5$, we determined that peak I ranging from fraction 1 to 25 contains $InsP_5$, peak II ranging from fraction 26 to 100 contains $InsP_6$, and peak III (fractions 100–160) contains diphosphoinositol pentakisphosphates. Due to the retention property of peaks IV and V and the elution after $PPInsP_5$ isomers upon MDD-HPLC analysis (Fig. 4), we assume that peak IV (fractions 160–224) contains bis-diphosphoinositol tetrakisphosphates, whereas peak V (fractions 224–280) contains tris-diphosphoinositol trisphosphates or more complex $InsP_9$ structures. In total, the solid-phase synthesis converts approximately 50% of $InsP_6$ to higher phosphorylated derivatives, about 26% to $PPInsP_5$, 17% to bis$PPInsP_4$, and 7% to tris$PPInsP_3$, respectively. Part of $InsP_6$ (17%) is dephosphorylated to a mixture of $InsP_5$ isomers. By a quantitative analytical parameterization of the reactions having occurred, we could show that formation of PPP-$InsP_5$ is probably impossible in our synthesis condition (see Note 2).

3.3. HPLC Analysis

1. Micro-metal-dye detection HPLC (Micro-MDD-HPLC) is performed on an anion-exchange column MiniQ™ PC 3.2/3 (see Note 3) as described previously (1) with some modifications. A HPLC chromatography system equipped with two gradient pumps is employed along with a Galaxie chromatographic workstation (Version 1.9) for instrument controlling as well as data acquisition and processing. Separation of $PP_xInsP_{(6-x)}$ is due to HCl gradient elution from the column, and detection by a SPD-10Avvp absorbance detector at 546 nm (or at 520 nm if possible) is brought about by postcolumn complexometric reaction with a detection solution C of 300 µM PAR in 1.6 M TEA (pH 9) (added in 1:2 flow ratio of reagent to eluent).

2. Eluents used in the gradient are: (a) 0.2 mM HCl and 15 µM YCl_3 and (b) 0.5 M HCl and 15 µM YCl_3. The presence of YCl_3 does not lead to complexation with phospho-compounds unless the pH is brought to over 4.5 by buffer.

3. Protocol of gradient I for separation of inositol phosphate isomers from $InsP_2$ to $InsP_8$ based on our commonly used method: 0–2 min, 3% B; 2.9–3.6 min, 5–7% B; 4.1–7.4 min, 9–10% B; 7.7–7.9 min, 11–13% B; 8.2–8.6 min, 15–17% B; 9.2–10.2 min, 18–19% B; 11.4–11.9 min, 24–28% B; 12.7–13.5 min, 35–45% B; 14.9–20 min, 58–65% B; 23.5–24 min, 80–100% B; 28–28.1 min, 100–3% B; and

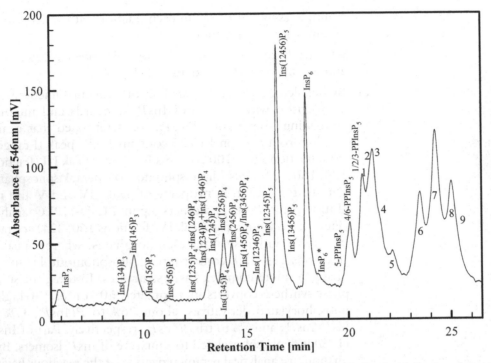

Fig. 6. MDD-HPLC analysis of an inositol phosphates mixture containing InsP$_2$ to InsP$_8$ (Gradient I). Numbers 1–9 indicate the nine bisPPInsP_4 isomers.

28.1–32 min, 3% B. Figure 6 shows the chromatogram where 27 isomers ranging from InsP$_2$ to InsP$_8$ are separated by using a standard mixture of reference compounds.

4. To optimize the separation of individual isomers of both PPInsP_5 and of bisPPInsP_4 without significant loss of sensitivity, two different gradients are applied.

5. Protocol of gradient II, optimized for separation of PPInsP_5 isomers: 0–4.7 min, 0–4% B; 5.6–7.6 min, 4–5% B; 7.6–11 min, 5–55% B; 11–28 min, 55–70% B; 28–31 min, 70–100% B; 31–32 min, 100–0% B; and 32–36 min, 0% B.

6. Protocol of gradient III, optimized for separation of bisPPInsP_4 isomers: 0–2 min, 3–45% B; 2–2.5 min, 45–58% B; 2.5–2.8 min, 58–59% B; 2.8–9 min, 59–65% B; 9–12.5 min, 65–80% B; 12.5–13 min, 80–100% B; 13–16 min, 100% B; 16–16.1 min, 100–3% B; and 16.1–20 min, 3% B.

7. The flow rates of the eluent and postcolumn reaction solution are 0.5 and 0.25 mL/min, respectively.

8. A knitted coil (75 μL dead volume) for postcolumn reaction is used.

9. HPLC performance is always at a stable temperature below 25°C to avoid the precipitation of PAR solution.

10. The injection volume of standard and sample solutions is always 1 mL by using an auto-sampler (Kontron AS560).

11. Sample loading of 1.2 mL is from 2-mL Pyrex glass vials where sample solution is always adjusted precisely to 1.3 mL volume by diluting with HPLC injection solution (2 mM sodium acetate, 2 mM NaF).

12. Prior to the chromatographic analysis, sample solutions are filtered through 0.20-µm Ministar RC4 membrane filters (Sartorius, Hannover, Germany).

13. Absorbance data, read continuously at 546 nm (or at 520 nm if possible) by a UV/VIS recorder (SPD-10Avvp, Shimadzu), are stored in polarity changed form (= multiplied with −1) and further processed as ASCII files with the Galaxie software (Varian), and SigmaPlot.

14. A representative HPLC analysis of selected fraction numbers 168, 182, 190, 200, and 220 from the chromatographic separation of shotgun-synthesized diphosphate inositol phosphates is shown in Fig. 5, where $InsP_6$ eluted at a retention time of 12.6 min. Using the isomeric forms of $PPInsP_5$ from Dr. Falck's laboratory (UT Southwestern, Dallas, USA) as external and internal standards revealed that 5-$PPInsP_5$ is the first isomer to be eluted from the HPLC column after $InsP_6$ at a retention time of 16.1 min, closely followed by 4- and 6-$PPInsP_5$ at a retention time of 16.5 min. The next peak eluted is as a mixture of 1/3-$PPInsP_5$ and 2-$PPInsP_5$ at a retention time of 17.7 min. A total of nine peaks are eluted after the $PPInsP_5$ isomers with retention times ranging from 18 to 28 min (Fig. 5), which we defined as our self-made chemically synthesized reference isomers $InsP_8$ #1 to #9. By pooling fractions or partly re-chromatographying this pooled material, we obtained chemically synthesized reference standards from all nine separable bis$PPInsP_4$ isomers. Their configuration was determined by further chemical and MDD-HPLC investigation and comparison with structurally characterized reference compounds.

3.4. Characterization of Nonenzymatic Synthesized Bis-diphosphate Inositol Tetrakisphosphate

1. By using a 2D NMR and chemically characterized and purified 5,6-bis$PPInsP_4$ from *Dictyostelium* (11), peak #3 in Fig. 7a can be attributed to this isomer (= $InsP_8$ #3). 5,6-bis$PPInsP_4$ co-chromatographed with $InsP_8$ #3 with both elution gradients employed. Retention time is 19.0 with gradient II (Fig. 7a).

2. As shown in Fig. 7b, a 2D NMR-characterized and purified 1/3,5-bis$PPInsP_4$ isomer from *Polysphondylium* (2 nmol) (12), used as a standard for HPLC co-chromatography analysis, has exactly the same elution property as $InsP_8$ #7 of our in-house $InsP_8$ reference (profile *b* and profile *c* in Fig. 7b).

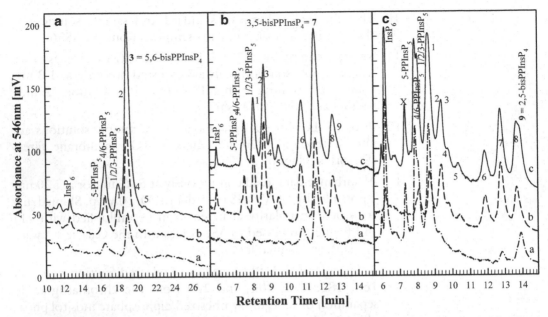

Fig. 7. Identification of bis*PP*Ins*P*$_4$ isomers by HPLC co-chromatography: (**a**) HPLC chromatograms of (profile *a*) *Dictyostelium* 5,6-bis*PP*Ins*P*$_4$, (profile *b*) chemically synthesized bis*PP*Ins*P*$_4$, InsP$_8$ #3 or (profile *c*) co-chromatography of a combination of bis*PP*Ins*P*$_4$ preparations (*a*) and (*b*) (Gradient II). (**b**) HPLC chromatograms of (profile *a*) 1/3,5-bis*PP* Ins*P*$_4$ standard (profile *b*) chemically synthesized bis*PP*Ins*P*$_4$, InsP$_8$ #7 or (profile *c*) co-chromatography of preparations (*a*) and (*b*) (Gradient III); and (**c**) HPLC chromatograms of (profile *a*) 2,5-bis*PP*Ins*P*$_4$ standard (Flack lab), (profile *b*) our chemically synthesized bis*PP*Ins*P*$_4$, InsP$_8$ #9 or (profile *c*) co-chromatography of preparations (*a*) and (*b*) (Gradient III).

The profile *a* in Fig. 7b also contains smaller amounts of InsP$_8$ #6, 5-*PP*Ins*P*$_5$, and 1/3-*PP*Ins*P*$_5$.

3. Furthermore, a position specifically synthesized 2,5-bis*P*-*PP*Ins*P*$_4$ (a gift from Prof. Falck, UT Southwestern, Dallas, USA) was analyzed by MDD-HPLC (profile *a* in Fig. 7c). As shown in Fig. 7c, this synthesized 2,5-bis*PP*Ins*P*$_4$, although heavily degraded, exhibited its major separable InsP$_8$ peak an elution property identical to InsP$_8$ #9 of our in-house InsP$_8$ reference (profile *c* in Fig. 7c).

4. Up to now, three isomers of bis*PP*Ins*P*$_4$ can be definitively (#3 and #7) or tentatively (#9) assigned, namely, #3 to 5,6- or 4,5-bis*PP*Ins*P*$_4$; #7 to 1,5- or 3,5-bis*PP*Ins*P*$_4$, and #9 to 2,5-bis*PP*Ins*P*$_4$.

3.5. Structure Resolution by Partial Hydrolysis of Bis-diphosphoinositol Tetrakisphosphate to PPInsP$_5$ Isomers

Based on the fact that the β-phosphates can be easily cleaved from the diphosphate groups of bis*PP*Ins*P*$_4$ by boiling in acid (8), we can use partial hydrolysis and isomeric identification of the resulting *PP*Ins*P*$_5$ hydrolytic intermediates to further proof the structure assignments. The chemically synthesized IP$_8$ #7 is further purified in large amounts by HPLC separation (profile *b* in Fig. 8) and used for partial hydrolysis.

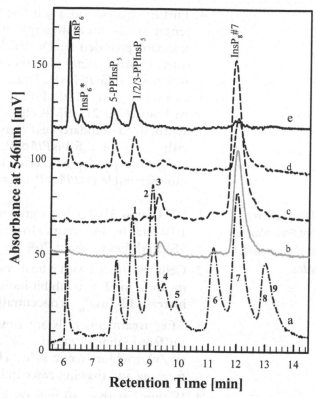

Fig. 8. HPLC chromatogram of a partial hydrolysate of nonenzymatically synthesized bis*PP*Ins*P$_4$*, Ins*P$_8$* #7. Profile *a*: Ins*P$_8$* standard isomeric mixture; profile *b*: purified Ins*P$_8$* #7 standard; profile *c*: products after 1.25 min TCA hydrolysis of purified Ins*P$_8$* #7; profile *d*: products after 10 min TCA hydrolysis of purified Ins*P$_8$* #7; profile *e*: products after 30 min TCA hydrolysis of purified Ins*P$_8$* #7. Ins*P$_6$** is likely a product where one phosphate ester and no pyrophosphate has been cleaved.

1. Transfer 25 μL of InsP_8 #7 solution (about 3–5 nmol) into a PTFE vial with a cap, add HCl or trichloroacetic acid to a final concentration of 0.5 M. Cap the vial tightly, and heat the solution at 95°C for between 1 and 30 min.

2. Cool down to room temperature, adjust pH to 6 with NaOH after HCl hydrolysis or triethanolamine base after trichloroacetic acid hydrolysis, and dilute with water for HPLC analysis.

3. Trichloroacetic acid treatment of the InsP_8 #7 for 15–30 min results in maximal formation of *PP*InsP_5 intermediates before they are further hydrolyzed. These *PP*InsP_5 intermediates co-eluted precisely with 5-*PP*InsP_5 and 1/3-or 2-*PP*InsP_5 isomers (profile *d* and *e* in Fig. 8); the position of 4/6-*PP*InsP_5 is in between and no such isomer was detected. We therefore conclude that the isomer InsP_8 #7 originally contained pyrophosphate groups at position 5 and 1/3 or 2.

4. Further analysis on a self-filled Mini Q column of 10 cm length by co-chromatography with 1-$PPInsP_5$ and 2-$PPInsP_5$ standard provided by Dr. Falck showed that the hydrolysis intermediate is identical to the slower eluting $PPInsP_5$ isomer, which is 1- or 3-$PPInsP_5$. Thus, we could prove only by limited hydrolysis and MDD-HPLC co-chromatography analysis without using 2D NMR-characterized or isomer specifically synthesized standard bis$PPInsP_4$ isomers that InsP_8 #7 is either 1,5- or 3,5-bis$PPInsP_4$. By this technique, we can assign the precise isomeric configuration to most of the nine separable bis$PPInsP_4$ isomers.

3.6. Cell Extraction of Inositol Phosphates and Diphosphoinositol Phosphates

1. NIH-Swiss 3 T3 cells are grown in DMEM (Invitrogen) with 10% fetal bovine serum (Invitrogen) in 5% CO_2 at 37°C on 15 cm dishes to about 70% confluence.

2. Cells are treated with a final concentration of 10 mM NaF in medium for 2 h to inhibit inositol pyrophosphatase and thus increase $PP_xInsP_{(6-x)}$ concentrations (10).

3. After treatment, cells are rinsed with ice-cold phosphate-buffered saline, scraped into ice-cold lysis buffer (1 mL 8% (w/v) trichloroacetic acid, 10 mM EDTA), followed by freezing and thawing twice in liquid nitrogen.

4. Within less than 30 min on ice, the precipitate is removed using centrifugation (3,500×g, 10 min, 4°C) and the supernatants extracted three times with 3 mL of ice-cold water-saturated diethyl ether.

5. After the pH has been adjusted to about 6 with 1 M TEA, the extract is then concentrated in a SpeedVac centrifuge.

6. The final extract is analyzed by MDD-HPLC (see Fig. 9).

7. A 2 h treatment with 10 mM fluoride of NIH 3 T3 cells resulted in the accumulation of large amounts of $PPInsP_5$ and bis$PPInsP_4$ (Fig. 9, profile c). The only major bis$PPInsP_4$ isomer detected has the same retention time as the isomer InsP_8 #7 and is identified as 1/3,5-bis$PPInsP_4$. Consistent with previous report (10), 5-$PPInsP_5$ is the major constituent of the $PPInsP_5$ isomer in mammalian cells (Fig. 9).

4. Notes

1. The column is preconditioned by washing with 10 column volume (CV) H_2O (2.4 ml in our case) at a flow rate of 0.5 mL/min, then with 10 CV of start buffer and elution buffer respectively. After equilibrium of the column with 10 CV start buffer, sample can be loaded with a flow rate of 0.5 mL/min.

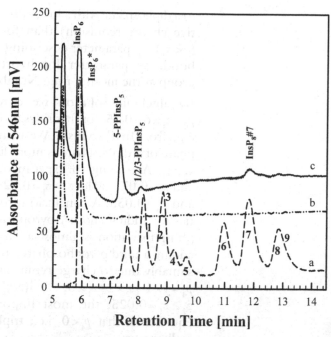

Fig. 9. HPLC chromatogram of cellular bis-diphosphoinositol tetrakisphosphate from NIH3T3 cells treated with 10 mM fluoride for 2 h (profile *c*) and from untreated NIH3T3 cells (profile *b*). Profile *a* is from InsP$_8$ standard isomeric mixture.

To start a new run, clean the column with 5 CV of elution buffer, before equilibrating with 5 CV of start buffer. After final run, the column is washed with 50-mL H$_2$O, and filled with 15% isopropanol in H$_2$O for storage.

2. Probabilities of nonenzymatic PP-group formation from InsP_6:

PPInsP_5:	$p_1 = x_1 \times 6 - z_2 \times 1 - z_1 \times 5 - x_2 \times 5$	(Fraction formed: 26%)
bisPPInsP_4:	$p_2 = x_2 \times 5 - z_2 \times 2 - z_1 \times 4 - x_3 \times 4$	(Fraction formed: 17%)
trisPPInsP_3:	$p_3 = x_3 \times 4 - z_2 \times 3 - z_1 \times 3$	(Fraction formed: 7%)
PPPInsP_5:	$p_4 = y \times 1 - z_2 \times 2 - z_1 \times 5$	(Fraction formed: ~0%)
InsP_5:	$p_0 = z_1 \times 6$	(Fraction formed: 15%)

No true rate constants could be deduced due to lack of real kinetic data, but from the fractions of isomeric mixtures of InsP_5, InsP_6, PPInsP_5, bisPPInsP_4, trisPPInsP_3, and PPPInsP_5 formed by the improved solid-phase shotgun synthesis technique on the Q-Sepharose beads, probabilities of formation can be parameterized. Hereby, the parameters setting up the formation probabilities are: x_n: parameter describing formation of diphosphate group depending on mass action, activation energy, and charge repulsion. y: parameter describing P-transfer to a diphosphate with a lower probability of attack (more negative charge repulsion) than for P-transfer to a phosphate ($y < x_1$). z_2: parameter describing hydrolysis of one anhydride bond. z_1: parameter for hydrolysis of one phosphoester group at the inositol ring. Numbers: stoichiometric factors.

By algebraic solution we determined: $z_1 = 0.025$ where $p_0 = z_1 \times 6 = 0.15$ for hydrolysis of one phosphoester group, e.g., from $InsP_6$ to $InsP_5$. We assume that hydrolysis of β-phosphate of a diphosphate is more likely than of a phosphoester, $z_2 \geq z_1$. According to experimental data, $p_1 = 0.26$, $p_2 = 0.17$, and $p_3 = 0.07$, and let $z_2 = z_1 = 0.025$, then $x_1 \geq 0.121$, $x_2 \geq 0.108$, and $x_3 \geq 0.055$. A conclusion from comparing the probabilities for adding another pyrophosphate is that a second pyrophosphorylation is almost as likely as the first one, whereas adding a third pyrophosphate group is only half as likely, presumably due to charge repulsion. For $PPPInsP_5$ formation: $p_4 = y \times 1 - z_2 \times 2 - z_1 \times 5$, hereby $y < x_1 \leq 0.121$ and $z_2 \geq z_1 = 0.025$; the most important conclusion from this equation is that $p_4 < 0$, i.e., triphosphate formation will not really occur.

3. Column performance worsens after about 50 runs. However, the column can be cleaned and performance fully restored according to the instructions by the manufacturer, namely by washing the column in reverse flow direction at a rate of 0.1 ml/min with 1 M NaCl for 10 min, then rinsing with water for 5 min; washing the column with 1 M NaOH for 10 min, rinsing with water for 5 min; washing the column with 1 M HCl for 10 min, then rinsing with water for 5 min; and with 1 M NaCl for 10 min and rinsing with water for 5 min. We used to redo the loop for 4 times to ensure the regeneration of column performance.

Acknowledgments

The authors would like to thank Professor Gunter Vogel (Wuppertal, Germany) for providing the 1/3,5-bis $PPInsP_4$ isomer from *Polysphondylium*, and Professor J.R. Falck (UT Southwestern, Dallas, USA) for providing a synthetic 2,5-bis $PPInsP_4$ sample and pure $PPInsP_5$ isomers. Parts of the MDD-HPLC analysis have been performed by Bettina Serreck, whose technical support is highly acknowledged.

References

1. Mayr, G.W. (1988) A novel metal-dye detection system permits picomolar-range h.p.l.c. analysis of inositol polyphosphates from non-radioactively labelled cell or tissue specimens. *Biochem. J.* **254**, 585–591.

2. Pittet, D., Schlegel, W., Lew, D., Monod, A., and Mayr, G. (1989) Mass changes in inositol tetrakis- and pentakisphosphate isomers induced by chemotactic peptide stimulation in HL-60 cells. *J. Biol. Chem.* **264**, 18489–18493.

3. Lorke, D.E., Gustke, H., and Mayr, G.W. (2004) An optimized fixation and extraction technique for high resolution of inositol phosphate signals in rodent brain. *Neurochem. Res.* **29**, 1887–1896.

4. Martin, J.B., Foray, M.F., Klein, G., and Satre, M. (1987) Identification of inositol hexaphosphate in ^{31}P-NMR spectra of *Dictyostelium discoideum* amoebae. Relevance to intracellular pH determination. *Biochim. Biophys. Acta.* **931**, 16–25.

5. Safrany, S.T., Caffrey, J.J., Yang, X., and Shears, S.B. (1999) Diphosphoinositol polyphosphates: the final frontier for inositide research? *Biol. Chem.* **380**, 945–951.

6. Menniti, F., Miller, R., Putney, J., Jr, and Shears, S. (1993) Turnover of inositol polyphosphate pyrophosphates in pancreatoma cells. *J. Biol. Chem.* **268**, 3850–3856.

7. Europe-Finner, G.N., Gammon, B., Wood, C.A., and Newell, P.C. (1989) Inositol tris- and polyphosphate formation during chemotaxis of *Dictyostelium*. *J. Cell Sci.* **93**, 585–592.

8. Mayr, G.W., Radenberg, T., Thiel, U., Vogel, G., and Stephens, L.R. (1992) Phosphoinositol disphosphates: non-enzymic formation in vitro and occurence in vivo in the cellular slime mold *Dictyostelium*. *Carbohydr. Res.* **234**, 247–262.

9. Stephens, L., Radenberg, T., Thiel, U., Vogel, G., Khoo, K.H., Dell, A., Jackson, T.R., Hawkins, P.T., Mayr, G.W., Stephens, L.R., Stanley, A.F., Moore, T., Poyner, D.R., Morris, P.J., Hanley, M.R., Kay, R.R., Irvine, R.F., Laussmann, T., Eujen, R., Weisshuhn, C.M., Martin, J.B., Bakker-Grunwald, T., Klein, G., Reddy, K.M., Reddy, K.K., and Falck, J.R. (1993) The detection, purification, structural characterization, and metabolism of diphospho-inositol pentakisphosphate(s) and bisdiphosphoinositol tetrakisphosphate(s) myo-inositol pentakisphosphates. *J. Biol. Chem.* **268**, 4009–4015.

10. Albert, C., Safrany, S.T., Bembenek, M.E., Reddy, K.M., Reddy, K., Falck, J., Brocker, M., Shears, S.B., and Mayr, G.W. (1997) Biological variability in the structures of diphosphoinositol polyphosphates in *Dictyostelium discoideum* and mammalian cells. *Biochem. J.* **327**, 553–560.

11. Laussmann, T., Reddy, K.M., Reddy, K.K., Falck, J.R., and Vogel, G. (1997) Diphospho-myo-inositol phosphates from *Dictyostelium* identified as D-6-diphospho-myo-inositol pentakisphosphate and D-5,6-bisdiphospho-myo-inositol tetrakisphosphate. *Biochem. J.* **322**, 31–33.

12. Laussmann, T., Hansen, A., Reddy, K.M., Reddy, K.K., Falck, J.R., and Vogel, G. (1998) Diphospho-myo-inositol phosphates in Dictyostelium and Polysphondylium: identification of a new bisdiphospho-myo-inositol tetrakisphosphate. *FEBS Lett.* **426**, 145–150.

13. Martin, J.-B., Bakker-Grunwald, T., and Klein, G. (1993) ^{31}P-NMR analysis of *Entamoeba histolytica*. Occurrence of high amounts of two inositol phosphates. *Eur. J. Biochem.* **214**, 711–718.

14. Bennett, M., Onnebo, S.M., Azevedo, C., and Saiardi, A. (2006) Inositol pyrophosphates: metabolism and signaling. *Cell Mol. Life Sci.* **63**, 552–564.

15. York, J.D. (2006) Regulation of nuclear processes by inositol polyphosphates. *Biochim. Biophys. Acta.* **1761**, 552–559.

16. Irvine, R.F. (2006) Nuclear inositide signalling – expansion, structures and clarification. *Biochim. Biophys. Acta.* **1761**, 505–508.

17. Luo, H.R., Saiardi, A., Yu, H., Nagata, E., Ye, K., and Snyder, S.H. (2002) Inositol pyrophosphates are required for DNA hyperrecombination in protein kinase C1 mutant yeast. *Biochemistry.* **41**, 2509–2515.

18. Luo, H.R., Huang, Y.E., Chen, J.C., Saiardi, A., Iijima, M., Ye, K., Huang, Y., Nagata, E., Devreotes, P., and Snyder, S.H. (2003) Inositol pyrophosphates mediate chemotaxis in *Dictyostelium* via pleckstrin homology domain-PtdIns(3,4,5)P$_3$ interactions. *Cell.* **114**, 559–572.

19. York, S.J., Armbruster, B.N., Greenwell, P., Petes, T.D., and York, J.D. (2005) Inositol diphosphate signaling regulates telomere length. *J. Biol. Chem.* **280**, 4264–4269.

20. Saiardi, A., Resnick, A.C., Snowman, A.M., Wendland, B., Snyder, S.H., Bhandari, R., Pesesse, X., Choi, K., Zhang, T., Shears, S.B., Luo, H.R., Huang, Y.E., Chen, J.C., Iijima, M., Ye, K., Huang, Y., Nagata, E., Devreotes, P., El Alami, M., Messenguy, F., Scherens, B., Dubois, E., Sciambi, C., McCaffery, J.M., Yu, H., Menniti, F.S., Miller, R.N., and Putney, J.W., Jr. (2005) Inositol pyrophosphates regulate cell death and telomere length through

phosphoinositide 3-kinase-related protein kinases phosphorylation of proteins by inositol pyrophosphates. *Proc. Natl. Acad. Sci. USA* **102**, 1911–1914.

21. York, J.D., Odom, A.R., Murphy, R., Ives, E.B., and Wente, S.R. (1999) A phospholipase C-dependent inositol polyphosphate kinase pathway required for efficient messenger RNA export. *Science.* **285**, 96–100.

22. Illies C, Gromada J, Fiume R, Leibiger B, Yu J, Juhl K, Yang SN, Barma DK, Falck JR, Saiardi A, Barker CJ, Berggren PO. (2007) Requirement of inositol pyrophosphates for full exocytotic capacity in pancreatic beta cells. *Science.* **318**, 1299–1302.

23. Saiardi, A., Bhandari, R., Resnick, A.C., Snowman, A.M., and Snyder, S.H. (2004) Phosphorylation of proteins by inositol pyrophosphates. *Science.* **306**, 2101–2105.

24. Choi, K., Mollapour, E., and Shears, S.B. (2005) Signal transduction during environmental stress: InsP(8) operates within highly restricted contexts. *Cell. Signal.* **17**, 1533–1541.

25. Pesesse, X., Choi, K., Zhang, T., and Shears, S.B. (2004) Signaling by higher inositolpolyphosphates: hyperosmotic stress acutely and selectively activates synthesis of bis-diphosphoinositol tetrakisphosphate ("InsP8"). *J. Biol. Chem.* **279**, 43378–43381.

26. Draskovic, P., Saiardi, A., Bhandari, R., Burton, A., Ilc, G., Kovacevic, M., Snyder, S.H., and Podobnik, M. (2008) Inositol hexakisphosphate kinase products contain diphosphate and triphosphate groups. *Chem. Biol.* **15**, 274–286.

27. Segel, I.H. (1976) Biochemical calculations: how to solve mathematical problems in general biochemistry. 2nd ed. New York: John Wiley & Sons Inc., pp. 15.

28. Lide, D.R. (2006) CRC Handbook Chemistry and Physics. 87th ed: Taylor & Francis Group, New York, pp. 8–41.

29. Lanzetta, P.A., Alvarez, L.J., Reinach, P.S., and Candia, O.A. (1979) An improved assay for nanomole amounts of inorganic phosphate. *Anal. Biochem.* **100**, 95–97.

Chapter 8

Diphosphosinositol Polyphosphates and Energy Metabolism: Assay for ATP/ADP Ratio

Andreas Nagel, Christopher J. Barker, Per-Olof Berggren, and Christopher Illies

Abstract

Several inositol compounds undergo rapid cycles of phosphorylation and dephosphorylation. These cycles are dependent on ATP and energy metabolism. Therefore, interfering with the cellular energy metabolism can change the concentration of rapidly turning over inositols. Many pharmacological inhibitors, apart from their intended action, also affect the energy metabolism of the cells and lower ATP. This can unspecifically influence rapidly turning over inositol phosphates. Thus, the ATP concentration should be checked when reduced inositol phosphates are observed after application of pharmacological inhibitors.

A luminescence-based assay for the measurement of ATP and ADP is described. ATP is measured luminometrically using firefly luciferase. Detection of ADP is performed in a two-step enzymatic procedure: (1) The sample ATP is degraded to AMP and (2) ADP is phosphorylated to ATP, which can then be measured luminometrically. This method gives a better signal-to-noise ratio than other methods that do not degrade the sample ATP, but convert ADP directly to ATP and then measure the sum of ATP plus ADP.

Key words: ATP/ADP ratio, Firefly luciferase, Luminescence, ATP-sulfurylase, HIT-T15, Diphosphoinositol pentakisphosphate

1. Introduction

Phosphorylation plays an important role in the biochemistry of inositol compounds, because the sterically unique phosphorylation determines the specificity of inositol signaling. Some inositol compounds, for example, the diphosphoinositol phosphates (inositol pyrophosphates) (1, 2) and phosphatidylinositol 4,5-bisphosphate (PtdInsP_2) (3), undergo rapid cycles of phosphorylation and dephosphorylation. The phosphorylation potential of diphosphoinositol pentakisphosphate (InsP_7) has been calculated to be slightly higher than that of ADP (2).

Christopher J. Barker (ed.), *Inositol Phosphates and Lipids: Methods and Protocols*, Methods in Molecular Biology, vol. 645, DOI 10.1007/978-1-60327-175-2_8, © Humana press, a part of Springer Science+Business Media, LLC 2010

These cycles are dependent on ATP and are sensitive to energetic stress of cells.

Pharmacological inhibitors are commonly used tools to investigate signaling pathways, but often they unspecifically lower cellular ATP by interfering with energy metabolism. This can affect synthesis of many inositol phosphates. Effect on energy metabolism should be checked when lowered inositol phosphate concentrations are seen after application of pharmacological agents.

The adenine nucleotide system, consisting of ATP, ADP, and AMP, is central to cellular energy metabolism as it couples the free energy released in catabolic reactions to anabolic processes. In these reactions, the adenine nucleotides are converted by addition or removal of a phosphate, rather than consumed or produced. Thus, a ratio of the adenine nucleotides, either the ATP/ADP ratio or the energy charge (4), provides a better index of the cellular energy state than just an absolute measurement of ATP. A ratiometric assay also provides an internal standard.

Firefly (*Photinus pyralis*) luciferase catalyzes the oxidation of luciferin using ATP and O_2 to produce light (5):

$$Luciferin + MgATP \leftrightarrow Luciferin\text{–}AMP + MgPP_i$$

$$Luciferin\text{–}AMP + O_2 \rightarrow oxyluciferin + CO2 + AMP + light$$

Firefly luciferase was used to measure adenine nucleotides (6) only a few years after the involvement of ATP in luminescence reactions had been described (7). The reaction is highly specific to ATP (8). To detect the other adenine nucleotides, these have to be converted to ATP. The commonly used method for measuring ADP and AMP is to measure ATP plus the sums ATP + ADP and ATP + ADP + AMP and then subtract the results (9–11).

ADP and AMP are usually only a small fraction of ATP, so when measuring these nucleotides as the sum of ATP + ADP or ATP + ADP + AMP, one is measuring a small signal on top of a high background. A way to avoid this problem is to degrade the ATP before the conversion of ADP to ATP (12). This is the method that will be described in this chapter. The detection of ADP is performed in the following steps: First the sample ATP is enzymatically degraded to AMP, and then ADP is quantitatively converted to ATP, which can be measured using firefly luciferase. The disadvantage of this assay is that AMP and the energy charge cannot be determined.

2. Materials

All reagents are obtained from Sigma-Aldrich, unless specified otherwise. Catalog numbers are given for enzymes.

2.1. Cell Lysis

1. 5% (w/v) trichloroacetic acid (TCA).
2. Water-saturated diethyl ether.
3. 0.1 M EDTA, pH 7.4.
4. 1 M triethanolamine (TEA); protect from light.
5. 12-well plate with HIT-T15 cells (5.5×10^5 cells/well, 1.5×10^5 cells/cm^2).
6. Krebs buffer: 119 mM NaCl, 20 mM HEPES–HCl pH 7.4, 4.6 mM KCl, 1 mM $MgSO_4$, 0.15 mM Na_2HPO_4, 0.4 mM KH_2PO_4, 5 mM $NaHCO_3$, 2 mM $CaCl_2$, 500 mg/L BSA, 0.1 mM glucose, pH 7.4.
7. Sodium azide.

2.2. Conversion of ATP to AMP

1. Adenosine-5′-triphosphate sulfurylase (ATP-sulfurylase) from *Saccharomyces cerevisiae* (A8957, Sigma): This is a lyophilised preparation containing ~40% protein with >1.0 U/mg protein. Prepare 100 U/mL stock solutions in water: 87% glycerol (1:1), aliquot, and store at –80°C.
2. 0.1 M Na_2MoO_4 stock.
3. 0.1 M GMP stock: Guanosine 5′-monophosphate disodium salt hydrate, aliquot and store at –20°C.
4. Other solutions: 1 M Tris–HCl pH 8.0, 0.1 M $MgCl_2$.

2.3. Conversion of ADP to ATP

1. Pyruvate kinase (10128155001, Roche Diagnostics), from rabbit muscle: 10 mg/mL suspension in 3.2 M $(NH_4)_2SO_4$.
2. 0.1 M phosphoenolpyruvate stock solution: Dissolve phosphoenolpyruvic acid tri(cyclohexylammonium) salt in water to form a 0.1-M solution, aliquot, and store at –80°C. The preparation keeps for several months.
3. 2× pyruvate kinase reaction buffer: 100 mM Tris–HCl pH 8.0, 5 mM $MgCl_2$, 40 mM KCl, 0.5 mM phosphoenolpyruvate, 0.1 mg/mL pyruvate kinase.

2.4. Detection of ATP

1. Luminescence plate reader (see Note 1).
2. Multi-well plates. Both 96-well plates and 384-well plates can be used. The plates must be opaque to reduce crosstalk. Manufacturers recommend white plates because of higher proton recovery. The volumes given in the text refer to 384-well plates (20 μl sample and 20 μl luciferase reagent), when 96-well plates are used 50 μl should be used for both sample and luciferase reagent.
3. 10 mM D-luciferin stock solution: Dissolve D-luciferin free base in water (N$_2$ bubbled) containing 10 mM $NaHCO_3$. Add NaOH until solution turns yellow (about 3 μl 0.4 M NaOH for 1 mg of D-luciferin). Aliquot and store at –80°C under N$_2$.

4. 50 mM coenzyme A (CoA) stock solution: Dissolve in 200 mM DTT, aliquot and store at $-20°C$ under N_2.

5. Luciferase, recombinant (L1792, Sigma): This preparation comes in a buffer ($\sim$5 mg protein/mL, $>5 \times 10^6$ light units/mg protein). Aliquot and store at $-80°C$; stable for at least 2 years.

6. Luciferase reagent buffer: 25 mM tricine pH 7.8, 5 mM $MgSO_4$, 0.1 mM EDTA, 0.05 mM D-luciferin, 2 μg/mL luciferase, 2 mM DTT, 0.05 mM CoA, 0.1 mg/mL BSA (see Note 2).

7. 2× sample buffer: 50 mM tricine pH 7.8, 10 mM $MgSO_4$, 0.2 mM EDTA.

8. ATP and ADP standards: For preparation of 100 mM ATP and ADP stock solutions, dissolve the sodium salts in water to about 70% of final volume. Adjust the pH to 7–8 with NaOH, check with pH paper. Determine the concentration spectro-photometrically (A_{259nm} (pH 7.0) = 15,400), and add water to a final concentration of 100 mM. Aliquot and store at $-80°C$.

3. Methods

The procedure for determining ATP/ADP ratios is as follows: First the cells are lysed and the adenine nucleotides are extracted. One part of the extract is saved for later measurement of ATP, whereas the part for ADP determination has to be processed further. It is first subjected to ATP-sulfurylase treatment to degrade ATP to AMP. Again, one part of this sample is saved to determine the leftover ATP, while the other is treated with pyruvate kinase to convert the ADP to ATP. The ADP concentration is calculated as the difference between the values obtained after pyruvate kinase and after ATP-sulfurylase treatment. The ATP concentration is determined directly from the extract.

In this example, the measurement of the ATP/ADP ratio is performed in HIT-T15 cells (13), cultured in 12-well plates and all the given volumes apply to that format. In these wells, the yield is usually about 5×10^{-9} mole ATP (2×10^{-5} mole ATP/g protein with 200 μg protein per well); the assay can easily be scaled up or down.

The cells are treated for 0 min, 5 min, and 30 min with the cytochrome c inhibitor sodium azide and the ATP/ADP ratio is determined for each condition. The ATP/ADP ratio is then compared to $InsP_7$ and $PtdInsP_2$ from similarly treated cells.

3.1. Cell Lysis

Several different methods for cell lysis and extraction of adenine nucleotides exist, including lysis in boiling buffer or acid extraction (see Note 3). Here, TCA was used. This method was demonstrated to give the highest yield of ATP (see Note 3) and it was also used for extraction of inositol phosphates. TCA is then removed by diethyl ether extraction.

1. Incubate one 12-well plate with HIT-T15 cells for 30 min in Krebs buffer. Then exchange buffer to Krebs buffer containing 0.05% sodium azide. Divide cells into four groups, each consisting of three replicates. Incubate for 0 min, 15 min, and 30 min.

2. Suck off culture medium, add 0.5 mL of 5% TCA solution to each well, and place the plate on ice for 10 min.

3. Scrape cells into a reaction tube, centrifuge and transfer supernatant to a fresh reaction tube.

4. Extract TCA four times with 1 mL diethyl ether: Add 1 mL of water-saturated diethyl ether, close tubes well, shake for 5 min, let stand until phases separate and suck off the upper phase. This should be performed under a fume hood. Ether is highly flammable and volatile, and appropriate care should be taken with handling and disposal.

5. Add 25 µl 0.1 M EDTA and adjust to pH 7–8 with TEA.

6. Take off remaining ether by drying samples in a centrifugal evaporator.

7. Store at –20°C.

3.2. Conversion of ATP to AMP

ATP is converted to AMP by the action of ATP-sulfurylase in the presence of molybdate (14). ATP-sulfurylase catalyzes the first step in the activation of sulfate:

$$MgATP + sulfate \rightarrow adenosine–5'–phosphosulfate + MgPP_i$$

If molybdate is used instead of sulfate, a stable adenylic acid-anion anhydride is not formed and ATP is cleaved to AMP and PP_i:

$$MgATP + molybdate \rightarrow AMP + MgPP_i + molybdate$$

Usually >99% of ATP is degraded to AMP. ATP and ADP standards should be run along with the samples (see Fig. 1) to control for the enzymatic procedures.

1. Resuspend dried samples in 500 µl of water.

2. Incubate samples in 200 µl of sulfurylase reaction buffer (see Note 4). Sulfurylase reaction buffer: 50 mM Tris–HCl pH 8.0, 5 mM $MgCl_2$, 10 mM Na_2MoO_4, 2.5 mM GMP, 30 U/mL ATP-sulfurylase.

3. Incubate for 1 h at 30°C.

4. Inactivate enzymes at 95°C for 2.5 min.

5. Spin down and put on ice.

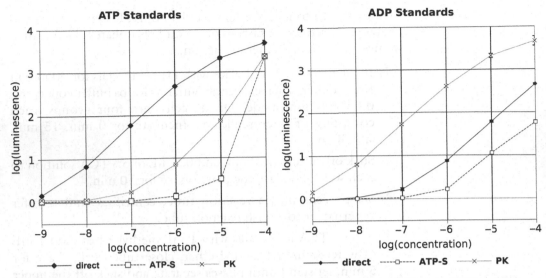

Fig. 1. ATP and ADP standard curves of a typical experiment. Standards ranging from 1 nM to 0.1 mM ATP and ADP were measured directly, after ATP-sulfurylase treatment (ATP-S), and after ATP-sulfurylase plus pyruvate kinase treatment (PK). The result was plotted as log(luminescence) vs. log(concentration). Note the incomplete elimination of ATP by ATP-sulfurylase at concentrations >1 μM. The contamination of ATP standards with ADP can be seen after pyruvate kinase treatment (see Note 5). Data points were determined in triplicate and plotted as mean ± standard deviation. Some error bars are covered by the symbols.

3.3. Conversion of ADP to ATP

The pyruvate kinase reaction is used to convert ADP to ATP. This reaction requires the presence of monovalent (K^+) and divalent (Mg^{2+}) cations (15).

$$ADP + phosphoenolpyruvate \rightarrow ATP + pyruvate$$

1. Take off 100 μl of ATP-sulfurylase reaction and add 100 μl of 2× pyruvate kinase reaction buffer.
2. Incubate for 1 h at room temperature.
3. Inactivate enzymes at 95°C for 5 min.
4. Add 800 μl of water.
5. Spin down and place on ice.

3.4. Detection of ATP

The luminescence reaction is started by adding luciferase buffer to the samples. The luminescence is stable for several minutes, because of the inclusion of CoA into the buffer (see Note 2). Thus, if no injector is available for the luminometer, luciferase buffer can be added manually to a whole plate just before it is read. The volumes and procedures given in the text refer to 384-well plates (20 μl sample and 20 μl luciferase reagent) using a plate reader with injection system. When 96-well plates are used 50 μl should be used for both sample and luciferase reagent.

1. Dilute ATP sample 1:200 with water.
2. Dilute all samples 1:2 with 2× sample buffer.

Table 1
ATP/ADP ratio in HIT-T15 cells after application of 0.05% sodium azide ($n=3$).
The ATP/ADP ratio was determined as described in this chapter. The inositol
phosphate data was from similarly treated HIT-T15 cells, which were labeled with
[^{3}H]-inositol for 120 h. Treatment with sodium azide strongly affects the ATP/ADP
ratio, even after 5 min. InsP_7, normalized against InsP_6, drops significantly as soon
as the ATP/ADP ratio goes down. PtdInsP_2 is also affected, but not as strongly
as InsP_7

Time (min)	ATP/ADP	Standard deviation	InsP_7 (% of InsP_6)	PtdInsP_2 (% of total lipids)
0	10.73	1.18	6.3	3.8
5	6.54	0.41	1.2	2.9
30	5.99	0.34	0.3	3.1

3. Transfer 20 µl per well of standards and samples to a 384-well plate. Use triplicates per data point.

4. Start luminescence reaction by injecting 20 µl of luciferase reagent.

5. Shake for 5 s to mix.

6. Measure luminescence.

3.5. Evaluation

ATP and ADP standards serve as control for the efficiency of the ATP-sulfurylase and pyruvate kinase reactions (see Fig. 1). To account for incomplete conversion of ADP to ATP, the ADP luminescence can be evaluated against the standard curve obtained from ADP after the pyruvate kinase reaction. The ATP remaining after ATP-sulfurylase treatment should be subtracted from that amount.

The effect of sodium azide on the ATP/ADP ratio is shown in Table 1. Even short incubation in azide reduces the ATP/ADP ratio strongly. The decrease of the ATP/ADP ratio correlates well to the decrease of InsP_7 in similarly treated cells. PtdInsP_2 decreases as well, but quickly reaches a plateau.

4. Notes

1. For a general discussion on plate readers for luminescence see ref. 16. In this lab, two different plate readers were used:

   (a) Wallac 1450 Microbeta, which could read up to 96-well plates and did not have an integrated injection and shaking system.

(b) 2103 Envision HTS Microplate Reader (PerkinElmer) equipped with Ultra Sensitive Luminescence and an injection system, which was used for assaying 384-well plates.

2. When different buffers were tested for their effect on luciferase activity, 25 mM tricine (pH 7.6) gave the highest light production (17) although Tris and Glycyl-glycine gave similar results, whereas HEPES and phosphate performed worse. A general discussion of the optimization of luciferase-based assays can be found in another volume of this series (16).

3. For quantitative recovery of the adenine nucleotides, the extraction method should provide full extraction of the adenine nucleotides and immediate inhibition of all adenine nucleotide converting enzymes. 1.25–10% TCA, the best concentration depending on the cell type, gives the highest recovery of adenine nucleotides when compared to other extraction methods (18). Because TCA interferes with further enzymatic reactions, it needs to be removed or diluted, before proceeding with the assay. TCA can then be removed by extraction with diethyl ether, followed by neutralization of the extract.

4. Nucleoside diphosphate kinase activity, which contaminates ATP-sulfurylase preparations, negatively interferes with the assay in the presence of GTP. Nucleoside diphosphate kinase uses the GTP to phosphorylate ADP to ATP, which in turn is degraded by ATP-sulfurylase. This leads to decreases of measured ADP. GMP serves to inhibit adenylate kinase activity, and higher GMP concentrations up to 15 mM might be needed when GTP levels are high (12). Contaminating activity appears to vary from batch to batch. Sometimes, addition of 6 mM sodium fluoride (19) is necessary for further inhibition of adenylate kinase. In this case, fluorolysis of ATP (20) seems to be less of a problem than contaminating adenylate kinase activity.

5. The ATP preparation was not completely pure, but contained about 1% of ADP and 1% of AMP when checked with HPLC/UV. The same is true for the ADP preparation, which contained about 1% of ATP. See also Fig. 1.

References

1. Menniti, F. S., Miller, R. N., Putney, J. W., and Shears, S. B. (1993) Turnover of inositol polyphosphate pyrophosphates in pancreatoma cells. *J. Biol. Chem.* **268**, 3850–3856.

2. Stephens, L., Radenberg, T., Thiel, U., Vogel, G., Khoo, K. H., Dell, A., Jackson, T. R., Hawkins, P. T., and Mayr, G. W. (1993) The detection, purification, structural characterization, and metabolism of diphosphoinositol pentakisphosphate(s) and bisdiphosphoinositol tetrakisphosphate(s). *J. Biol. Chem.* **268**, 4009–4015.

3. Poggioli, J., Weiss, S. J., McKinney, J. S., and Putney, J. W. (1983) Effects of antimycin a on receptor-activated calcium mobilization and phosphoinositide metabolism in rat parotid gland. *Mol. Pharmacol.* **23**, 71–77.

4. Atkinson, D. E. (1977) *Cellular energy metabolism and its regulation*. Academic Press, New York.

5. Viviani, V. R. (2002) The origin, diversity, and structure function relationships of insect luciferases. *Cell. Mol. Life Sci.* **59**, 1833–1850.

6. Strehler, B. L. and Totter, J. R. (1954) Determination of ATP and related compounds: firefly luminescence and other methods. *Methods Biochem. Anal.* **1**, 341–356.

7. McElroy, W. D. (1947) The energy source for bioluminescence in an isolated system. *Proc. Natl. Acad. Sci. U. S. A.* **33**, 342–345.

8. Moyer, J. D. and Henderson, J. F. (1983) Nucleoside triphosphate specificity of firefly luciferase. *Anal. Biochem.* **131**, 187–189.

9. Pradet, A. (1967) Étude des adénosine-5′-mono, de et triphosphates dans les tissues végétaux, i. Dosage enzymatique. *Physiol. Vég.* **5**, 209–221.

10. Lundin, A. and Thore, A. (1975) Comparison of methods for extraction of bacterial adenine nucleotides determined by firefly assay. *Appl. Microbiol.* **30**, 713–721.

11. Ford, S. R. and Leach, F. R. (1998) Bioluminescent assay of the adenylate energy charge. In: LaRossa, L. A., ed., *Methods in Molecular Biology*, vol. 102: Bioluminescence Methods and Protocols. Humana Press Inc., Totowa, NJ.

12. Schultz, V., Sussman, I., Bokvist, K., and Tornheim, K. (1993) Bioluminometric assay of ADP and ATP at high ATP/ADP ratios: assay of ADP after enzymatic removal of ATP. *Anal. Biochem.* **215**, 302–304.

13. Santerre, R. F., Cook, R. A., Crisel, R. M., Sharp, J. D., Schmidt, R. J., Williams, D. C., and

Wilson, C. P. (1981) Insulin synthesis in a clonal cell line of simian virus 40-transformed hamster pancreatic beta cells. *Proc. Natl. Acad. Sci. U. S. A.* **78**, 4339–4343.

14. Wilson, L. G. and Bandurski, R. S. (1958) Enzymatic reactions involving sulfate, sulfite, selenate, and molybdate. *J. Biol. Chem.* **233**, 975–981.

15. Kayne, F. J. (1973) Pyruvate kinase. In: Boyer, P. D., ed., *The Enzymes*, vol. 8A. Academic Press, New York, pp. 353–382, 3 edn.

16. Ford, S. R. and Leach, F. R. (1998) Improvements in the application of firefly luciferase assays. In: LaRossa, L. A., ed., *Methods in Molecular Biology*, vol. 102: Bioluminescence Methods and Protocols. Humana Press Inc., Totowa, NJ.

17. Webster, J. J., Chang, J. C., Manley, E. R., Spivey, H. O., and Leach, F. R. (1980) Buffer effects on ATP analysis by firefly luciferase. *Anal. Biochem.* **106**, 7–11.

18. Lundin, A., Hasenson, M., Persson, J., and Pousette, A. (1986) Estimation of biomass in growing cell lines by adenosine triphosphate assay. *Methods Enzymol.* **133**, 27–42.

19. Meiattini, F., Giannini, G., and Tarli, P. (1978) Adenylate kinase inhibition by adenosine 5-monophosphate and fluoride in the determination of creatine kinase activity. *Clin. Chem.* **24**, 498–501.

20. London, R. E. and Gabel, S. A. (1996) Mg^{2+} and other polyvalent cations catalyze nucleotide fluorolysis. *Arch. Biochem. Biophys.* **334**, 332–340.

Chapter 9

Isolation of Inositol 1,4,5-Trisphosphate Receptor-Associating Proteins and Selective Knockdown Using RNA Interference

Akihiro Mizutani, Katsuhiro Kawaai, Chihiro Hisatsune, Hideaki Ando, Takayuki Michikawa, and Katsuhiko Mikoshiba

Abstract

Inositol 1,4,5-trisphosphate (IP$_3$) receptors (IP$_3$Rs) are IP$_3$-gated Ca^{2+} release channels localized on intracellular Ca^{2+} stores and play a role in the generation of complex patterns of intracellular Ca^{2+} signals. We show herein experimental protocols for the identification of associating proteins of IP$_3$R isoforms from various cells and tissues using affinity column chromatography and for the specific knockdown of the expression of IP$_3$R isoforms and their associating proteins using RNA interference. These methods will provide clues to understand the exact nature of how the signaling complex contributes to the generation of spatio-temporal patterns of intracellular Ca^{2+} signals.

Key words: IP$_3$, Ca^{2+}, Affinity chromatography, GST-fusion protein, siRNA, Ca^{2+} oscillation, IRBIT, ERp44

1. Introduction

Inositol 1,4,5-trisphosphate (IP$_3$) receptors (IP$_3$Rs) are IP$_3$-gated channels that release Ca^{2+} from the endoplasmic reticulum in response to an elevation of cytoplasmic IP$_3$ concentration and play a critical role in establishing diverse patterns of intracellular Ca^{2+} signals (1–3). The complex spatial and temporal patterns of Ca^{2+} signals, such as Ca^{2+} waves and Ca^{2+} oscillations, regulate many cellular responses, including fertilization, muscle contraction, secretion, cell growth, differentiation, apoptosis, and synaptic plasticity (4). The diversity of Ca^{2+} signal generated in a spatial and temporal fashion may be attributable to the existence of three

Christopher J. Barker (ed.), *Inositol Phosphates and Lipids: Methods and Protocols*, Methods in Molecular Biology, vol. 645, DOI 10.1007/978-1-60327-175-2_9, © Humana press, a part of Springer Science+Business Media, LLC 2010

isoforms of IP$_3$Rs, IP$_3$R1, IP$_3$R2, and IP$_3$R3 that show distinct properties in terms of their IP$_3$ sensitivity (5), the modulatory effects on them of cytoplasmic Ca^{2+} (6), and their unique tissue distribution (7). Recent advance in the identification of IP$_3$R-binding proteins revealed that IP$_3$Rs form a signaling complex with their binding proteins, which modulate the channel activity or intracellular distribution of IP$_3$Rs (8), suggesting that the dynamic interaction between IP$_3$Rs and accessory proteins might also contribute to the generation of the diversity of Ca^{2+} signals.

To obtain a better understanding of the mechanism of the generation of complex patterns of Ca^{2+} signals, the effective combination of two distinct approaches, (1) the identification of proteins, which associate with IP$_3$R isoforms and (2) the knockdown of the expression of specific proteins in living cells, has considerable promise. We have successfully identified IP$_3$R-binding proteins, IRBIT (9) and ERp44 (10), from the mouse cerebellum using IP$_3$R1-affinity chromatography. Loss-of-function experiments with these associating proteins using RNA interference (RNAi), which is a rapid and convenient experimental technique compared with the conventional gene targeting technique, clearly demonstrated that both proteins provide negative regulation of the IP$_3$R function in living cells (10, 11). We have also established a method for the isoform-specific knockdown of IP$_3$Rs using RNAi and found that IP$_3$R1 and IP$_3$R3 contribute differently in the generation of temporal patterns of Ca^{2+} oscillations in cultured cells (12) (Fig. 1). We show herein the experimental protocols for the identification of IP$_3$R-associating proteins using IP$_3$R-affinity column chromatography and for the knockdown of the expression of IP$_3$R isoforms and IP$_3$R-associating proteins using RNAi.

2. Materials

2.1. Purification of Accessory Proteins of IP$_3$R Isoforms

2.1.1. Preparation of IP$_3$R-Affinity Resins

1. Sf9 cells maintained in SF-900 II serum-free medium (Invitrogen).
2. DH10Bac™ competent cells (Invitrogen).
3. Bac-to-Bac® baculovirus expression kit (Invitrogen).
4. Sf9 cell homogenization buffer: 10 mM HEPES (pH 7.4), 100 mM NaCl, 2 mM EDTA, 1 mM 2-mercaptoethanol, 0.1% Triton X-100, and a cocktail of protease inhibitors: 1 mM phenylmethylsulfonyl fluoride (Sigma-Aldrich), 10 µM leupeptin (Peptide Institute), 2 µM pepstatin A (Peptide Institute), and 10 µM E-64 (Peptide Institute).
5. Wash buffer: 10 mM HEPES (pH 7.4), 250 mM NaCl, 2 mM EDTA, 1 mM 2-mercaptoethanol, and 0.1% Triton X-100.

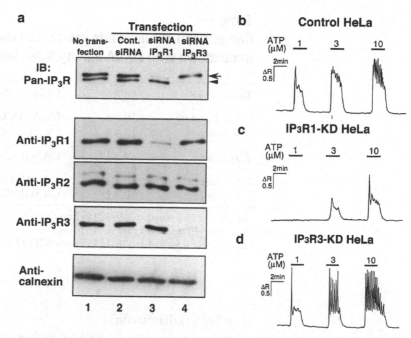

Fig. 1. Selective knockdown of IP_3R1 and IP_3R3 in HeLa cells and effects of IP_3R knockdown on ATP-induced Ca^{2+} dynamics. (**a**) Western blot analysis of HeLa cells transfected with the siRNAs indicated. Immunoblotting (IB) with the IP_3R isoform-specific antibodies, anti-IP_3R1, anti-IP_3R2, and anti-IP_3R3, was performed for cell lysates prepared 48 h after transfection. The pan-IP_3R antibody (Pan-IP_3R) recognizes all three IP_3R isoforms. (**b–d**) The results of single-cell Ca^{2+} imaging of control HeLa cells (**b**), IP_3R1-knockdown cells (**c**), and IP_3R3-knockdown cells (**d**). ATP-induced cytosolic Ca^{2+} concentration changes were monitored with Fura-2. Modified from Hattori et al. (12).

6. Glutathione-Sepharose 4B (GE Healthcare).

7. FLAG peptide (Sigma-Aldrich).

8. Anti-FLAG® M2 affinity gel (Sigma-Aldrich).

2.1.2. Preparation of Crude Lysates from Tissues or Cells

1. Homogenization buffer: 10 mM HEPES (pH 7.4), 320 mM sucrose, 2 mM EDTA, 1 mM 2-mercaptoethanol, and the cocktail of protease inhibitors.

2. Glutathione-Sepharose 4B (GE Healthcare).

2.1.3. Purification and Identification of IP₃R-Binding Proteins

1. Glutathione (Wako Chemical).

2. Elution buffer: 10 mM HEPES (pH 7.4), 0.05% Triton X-100, 100 mM NaCl, 2 mM EDTA, and 1 mM 2-mercaptoethanol.

2.2. Knockdown of IP₃Rs and Their Associated Proteins by RNAi

1. COS-7 and HeLa cells maintained in Dulbecco's modified essential medium (Nakarai Tesque) supplemented with 10% fetal bovine serum, 50 units/mL penicillin, and 50 μg/mL streptomycin (Nakarai Tesque).

Table 1
Comparison of nucleotide sequences used for selective knockdown with siRNA among three human IP$_3$R isoforms

Target gene	IP$_3$R1	UGAGACAGAAAACAGGAAA[a]
Corresponding sequence	IP$_3$R2	UGA**AT**CGGAGA**AT**AAGAAA[b]
	IP$_3$R3	UGA**CAC**GGAGA**AC**AAGAA**G**[b]
Target gene	IP$_3$R3	GAAGUUCCGUGACUGCCUC[c]
Corresponding sequence	IP$_3$R1	GAA**A**UUCAGAGACUGCCUC[b]
	IP$_3$R2	GAAGUUCAGAGACUGCCU**U**[b]

[a]The sequence is located at 309–327 in human IP$_3$R1 mRNA
[b]Nonconserved nucleotides are shown in *bold*
[c]The sequence is located at 150–168 in human IP$_3$R3 mRNA

2. siRNA (Dharmacon).
   IP$_3$R1: UGAGACAGAAAACAGGAAA (12) (Table 1).
   IP$_3$R3: GAAGUUCCGUGACUGCCUC (12) (Table 1).
   IRBIT: AACUCAGAAUGAAGUAGCUGC (11).
   ERp44: AAGUAGUGUUUGCCAGAGUUG (10).
3. Opti-MEM® and Lipofectamine™ 2000 (Invitrogen).
4. FuGENE® HD transfection reagent (Roche Applied Science).

3. Methods

3.1. Purification of Accessory Proteins of IP$_3$R Isoforms

The structure of IP$_3$Rs has traditionally been divided into three functional domains: the N-terminal IP$_3$-binding domain, the internal modulatory/coupling domain, and the C-terminal transmembrane/channel-forming domain (Fig. 2). We have developed a method for screening of proteins that are associated with the N-terminal cytoplasmic region, which includes both the N-terminal IP$_3$-binding domain and a large part of the internal modulatory/coupling domain, of three mouse IP$_3$R isoforms (5, 13). This method uses the N-terminal 2217, 2171, and 2145 amino acid residues of IP$_3$R1, IP$_3$R2, and IP$_3$R3, respectively, as a bait for affinity purification (Fig. 2) and has considerable advantages of flexibility of experimental conditions, such as concentrations of Ca^{2+}, IP$_3$, nucleotides, and so on, in the solutions used for both the absorption and elution steps. This method is also applicable for the screening of phosphorylation state-dependent associating proteins of IP$_3$R isoforms.

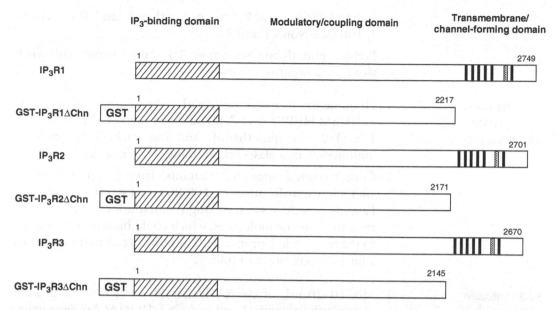

Fig. 2. Structures of IP$_3$R isoforms and constructs used for the production of IP$_3$R-affinity columns. The structure of IP$_3$Rs has been divided into three functional domains: the N-terminal IP$_3$-binding domain, the internal modulatory/coupling domain, and the C-terminal transmembrane/channel-forming domain. Six transmembrane segments (*thick black lines*) and one pore segment (*dotted boxes* between fifth and sixth transmembrane segments) are shown for each isoform. IP$_3$-binding domains are shown in diagonally *shaded boxes*. GST was fused to the N-terminus of the N-terminal cytosolic regions (1-2217 of IP$_3$R1, 1-2171 of IP$_3$R2, and 1-2145 of IP$_3$R3) of three mouse IP$_3$R isoforms (2749, 2701, and 2670 amino acid residues, respectively) for the preparation of the IP$_3$R-affinity columns.

3.1.1. Preparation
of IP$_3$R-Affinity Resins

1. Prepare recombinant baculoviruses encoding the glutathione S-transferase (GST)-fused N-terminal cytoplasmic regions of IP$_3$R isoforms, GST-IP$_3$R1ΔChn, GST-IP$_3$R2ΔChn, and GST-IP$_3$R3ΔChn (Fig. 2), with a Bac-to-Bac® baculovirus expression system (Invitrogen) and DH10Bac™ competent cells (Invitrogen) according to the manufacturer's protocols.

2. Infect Sf9 cells (100 mL of 2×10^8 cells) with recombinant baculoviruses in SF-900 II serum-free medium (Invitrogen) at a multiplicity of infection of 5. Incubate the infected cells in a 26°C shaking incubator for 48 h.

3. Homogenize the infected Sf9 cells in 10 mL of 10 mM HEPES (pH 7.4), 100 mM NaCl, 2 mM EDTA, 1 mM 2-mercaptoethanol, 0.1% Triton X-100, and a cocktail of protease inhibitors (1 mM phenylmethylsulfonyl fluoride, 10 μM leupeptin, 2 μM pepstatin A, and 10 μM E-64) by sonication. Collect supernatants after centrifugation of the homogenates at $20,000 \times g$ for 30 min at 4°C.

4. Incubate the supernatants with 3 mL of glutathione-Sepharose 4B resin (GE Healthcare) for 3 h at 4°C. Wash the resin with 50 mL of 10 mM HEPES (pH 7.4), 250 mM NaCl,

2 mM EDTA, 1 mM 2-mercaptoethanol, and 0.1% Triton X-100 (see Notes 1 and 2).

5. Prepare glutathione-Sepharose 4B resin treated with GST alone for a negative control.

3.1.2. Preparation of Crude Lysates from Tissues or Cells

1. Homogenize tissues or cells (with a net weight of at least 1 g) in 10 mM HEPES (pH 7.4), 320 mM sucrose, 2 mM EDTA, 1 mM 2-mercaptoethanol, and the cocktail of protease inhibitors with a glass-Teflon homogenizer on ice.

2. Collect crude lysates (supernatants) from the homogenates after ultracentrifugation at $100,000 \times g$ for 60 min at 4°C. Pass the crude lysates through glutathione-Sepharose 4B resin to eliminate molecules, which could bind nonspecifically to the resin. Add Triton X-100 into the precleared samples to a final concentration of 0.05%.

3.1.3. Purification and Identification of IP_3R-binding Proteins

1. Mix 10–20 mL of cleared lysates (1–5 mg/mL of protein concentration) with 1 mL of GST-IP_3R1ΔChn-Sepharose, GST-IP_3R2ΔChn-Sepharose, or GST-IP_3R3ΔChn-Sepharose. Adjust the free Ca^{2+} concentration of the mixture and/or add small compounds, such as IP_3, ATP, GTP, and so forth, in the mixture as required.

2. Incubate the mixture for more than 2 h at 4°C.

3. Wash the mixture with the solution having the same concentrations of free Ca^{2+}, IP_3, nucleotides, and so on as adjusted or added in step 1 (see Note 3).

4. Elute IP_3R-associating protein-IP_3R fragment complexes from Sepharose resin with 10 mM glutathione in 10 mM HEPES (pH 7.4), 0.05% Triton X-100, 100 mM NaCl, 2 mM EDTA, and 1 mM 2-mercaptoethanol (see Note 4).

5. Apply eluted proteins on SDS-PAGE and visualize IP_3R-associating proteins with Coomassie Brilliant Blue R-250 (CBB) staining or silver staining. Incubate the proteins with appropriate proteases in the gel and pass to mass spectrometric analysis for identification of the proteins isolated.

3.2. Knockdown of IP_3Rs and Their Associated Proteins by RNAi

RNAi is an RNA-dependent gene silencing process that is controlled by small interfering RNA strands (siRNA), which have complementary nucleotide sequences to the targeted RNA strand. RNAi plays a role in regulating development and genome maintenance, but it can be used for the analysis of the role of specific proteins in cell functions by the introduction of synthetic siRNAs into cells (14). Because siRNA can have nonspecific effects (15), careful control experiments are required for specific knockdown of targeted gene products.

1. Plate the cells (2.0 mL of 1×10^5 cells) in a 35-mm tissue-culture dish or glass-bottom dish. Incubate the cells for 24 h in a CO_2 incubator to achieve 60–80% confluency at the time of transfection.

2. Replace the medium with 2 mL of fresh culture medium 30 min prior to transfection.

3. Mix 5 µl of 20 µM of siRNA (see Note 5) and 250 µl of Opti-MEM® (Invitrogen) by gentle pipetting and/or tapping (see Note 6). Mix 5 µl of Lipofectamine™ 2000 (Invitrogen) (see Note 7) and 250 µl of Opti-MEM® in another tube.

4. Combine two mixtures gently and incubate it for 20 min at room temperature.

5. Add the entire mixture onto the cells to a final concentration of 40 nM of siRNA.

6. Incubate the cells for 5–8 h in a CO_2 incubator. Wash the cells once with phosphate-buffered saline and then add 2 mL of fresh culture medium. Incubate the cells for 48 h (see Note 8).

4. Notes

4.1. Purification of Accessory Proteins of IP$_3$R Isoforms

1. Approximately 5 mg of GST-IP$_3$R1ΔChn protein should be immobilized on 3 mL of Sepharose. The immobilized proteins can be quantified by applying the treated resin directly onto SDS-PAGE and measuring the intensity of the GST-IP$_3$R fragments after CBB staining. In the case of GST-IP$_3$P2ΔChn and GST-IP$_3$R3ΔChn, degradation products tend to be observed with much higher contents compared with the case of GST-IP$_3$R1ΔChn. To improve the yield of intact proteins immobilized on Sepharose resin, we added FLAG sequence to the C-terminus of GST-IP$_3$R2ΔChn and GST-IP$_3$R3ΔChn. Proteins with FLAG sequence eluted from anti-FLAG® M2 affinity gel (Sigma-Aldrich) with 100 µg/mL of FLAG peptide should be used for the reaction with glutathione-Sepharose 4B.

2. For the isolation of phosphorylation state-dependent associating proteins of IP$_3$R isoforms, GST-fusion proteins are eluted from glutathione-Sepharose resin with 10 mM glutathione. Eluted proteins can be subjected to phosphorylation with various protein kinases. The phosphorylated GST-fusion proteins should be recaptured on another glutathione-Sepharose resin and can be used for affinity purification.

3. It is worth checking what proteins were washed out in each washing cycle to avoid missing proteins with a weak affinity against IP$_3$Rs.

4. The elution of proteins may be tried by just changing the concentrations of compounds of interest, instead of by adding 10 mM of glutathione. We isolated IRBIT from mouse cerebellum by adding IP$_3$ in the solution used for elution from GST-IP$_3$R1ΔChn-Sepharose (9).

4.2. Knockdown of IP$_3$Rs and Their Associated Proteins by RNAi

5. Twenty micromoles of siRNA stock solution was prepared in 20 mM KCl, 6 mM HEPES–KOH (pH 7.5), 0.2 mM MgCl$_2$. It is desirable to use two or three point mutants of siRNA used for gene silencing as a negative control. Commercially available general-purpose negative control siRNAs with a similar GC content with that of siRNA designed for specific knockdown are also applicable. In both cases, the expression level should be checked not only of the target gene products but also the homeostatic gene products, such as β-actin and glyceraldehyde 3-phosphate dehydrogenase (GAPDH), to evaluate nonspecific effects of the siRNAs introduced. The expression level of gene products related to the target genes, such as proteins in a same gene family, should be analyzed to elicit the compensational effect of the knockdown of the target.

6. The amount of reagents described is just a starting point. To obtain the best results, optimize the balance of siRNA concentration and transfection reagent concentration for individual cell types and target genes.

7. Because Lipofectamine™ 2000 is slightly toxic, we also used FuGENE® HD transfection reagent (Roche Applied Science) for transfection of cells. We used 1.5 μg of siRNA and 8 μl of FuGENE® HD transfection reagent for COS-7 cells (see ref. 16 for more details).

8. Figure 1 shows the result of the selective knockdown of IP$_3$R1 and IP$_3$R3 in HeLa cells. If the knockdown efficiency of the targets is too low, both the exposure time of the transfection complex and the incubation time after transfection should be optimized. If the efficiency is still insufficient after the optimization, try different siRNAs designed for the target genes. Knockdown of IP$_3$R2 has been reported by other groups (17, 18).

Acknowledgments

This work was supported by grants from the Education, Culture, Sports, and Technology of Japan.

References

1. Bezprozvanny, I. (2005) The inositol 1,4,5-trisphosphate receptors. *Cell Calcium* **38**, 261–272.

2. Foskett, J. K., White, C., Cheung, K. H. and Mak, D. O. (2007) Inositol trisphosphate receptor Ca²⁺ release channels. *Physiol. Rev.* **87**, 593–658.

3. Mikoshiba, K. (2007) IP₃ receptor/Ca²⁺ channel: from discovery to new signaling concepts. *J. Neurochem.* **102**, 1426–1446.

4. Berridge, M. J. (1993) Inositol trisphosphate and calcium signalling. *Nature* **361**, 315–325.

5. Iwai, M., Tateishi, Y., Hattori, M., Mizutani, A., Nakamura, T., Futatsugi, A., Inoue, T., Furuichi, T., Michikawa, T. and Mikoshiba, K. (2005) Molecular cloning of mouse type 2 and type 3 inositol 1,4,5-trisphosphate receptors and identification of a novel type 2 receptor splice variant. *J. Biol. Chem.* **280**, 10305–10317.

6. Tu, H., Wang, Z., Nosyreva, E., De Smedt, H. and Bezprozvanny, I. (2005) Functional characterization of mammalian inositol 1,4,5-trisphosphate receptor isoforms. *Biophys. J.* **88**, 1046–1055.

7. Newton, C. L., Mignery, G. A. and Sudhof, T. C. (1994) Co-expression in vertebrate tissues and cell lines of multiple inositol 1,4,5-trisphosphate (InsP₃) receptors with distinct affinities for InsP₃. *J. Biol. Chem.* **269**, 28613–28619.

8. Choe, C. U. and Ehrlich, B. E. (2006) The inositol 1,4,5-trisphosphate receptor (IP₃R) and its regulators: sometimes good and sometimes bad teamwork. *Sci. STKE* **2006**, re15.

9. Ando, H., Mizutani, A., Matsu-ura, T. and Mikoshiba, K. (2003) IRBIT, a novel inositol 1,4,5-trisphosphate (IP₃) receptor-binding protein, is released from the IP₃ receptor upon IP₃ binding to the receptor. *J. Biol. Chem.* **278**, 10602–10612.

10. Higo, T., Hattori, M., Nakamura, T., Natsume, T., Michikawa, T. and Mikoshiba, K. (2005) Subtype-specific and ER lumenal environment-dependent regulation of inositol 1,4,5-trisphosphate receptor type 1 by ERp44. *Cell* **120**, 85–98.

11. Ando, H., Mizutani, A., Kiefer, H., Tsuzurugi, D., Michikawa, T. and Mikoshiba, K. (2006) IRBIT suppresses IP₃ receptor activity by competing with IP₃ for the common binding site on the IP₃ receptor. *Mol. Cell* **22**, 795–806.

12. Hattori, M., Suzuki, A. Z., Higo, T., Miyauchi, H., Michikawa, T., Nakamura, T., Inoue, T. and Mikoshiba, K. (2004) Distinct roles of inositol 1,4,5-trisphosphate receptor types 1 and 3 in Ca²⁺ signaling. *J. Biol. Chem.* **279**, 11967–11975.

13. Furuichi, T., Yoshikawa, S., Miyawaki, A., Wada, K., Maeda, N. and Mikoshiba, K. (1989) Primary structure and functional expression of the inositol 1,4,5-trisphosphate-binding protein P₄₀₀. *Nature* **342**, 32–38.

14. Dykxhoorn, D. M. and Lieberman, J. (2005) The silent revolution: RNA interference as basic biology, research tool, and therapeutic. *Ann. Rev. Med.* **56**, 401–423.

15. Scherer, L. J. and Rossi, J. J. (2003) Approaches for the sequence-specific knockdown of mRNA. *Nat. Biotechnol.* **21**, 1457–1465.

16. Kawaai, K., Ishida, S., Matsu-ura, T., Kuroda, Y., Ogawa, N., Tashiro, T., Enomoto, M., Hisatsune, C. and Mikoshiba, K. (2008) Optimization of transfection conditions for gene expression, siRNA knock-down, and live cell imaging using FuGene HD Transfection reagent. *Biochemica* **280**, 21–24.

17. Mendes, C. C., Gomes, D. A., Thompson, M., Souto, N. C., Goes, T. S., Goes, A. M., Rodrigues, M. A., Gomez, M. V., Nathanson, M. H. and Leite, M. F. (2005) The type III inositol 1,4,5-trisphosphate receptor preferentially transmits apoptotic Ca²⁺ signals into mitochondria. *J. Biol. Chem.* **280**, 40892–40900.

18. Galeotti, N., Quattrone, A., Vivoli, E., Bartolini, A. and Ghelardini, C. (2007) Knockdown of the type 2 and 3 inositol 1,4,5-trisphosphate receptors suppresses muscarinic antinociception in mice. *Neuroscience* **149**, 409–420.

Chapter 10

Phosphoinositide-Specific Phospholipase C β1 Signal Transduction in the Nucleus

Roberta Fiume, Gabriella Teti, Irene Faenza, and Lucio Cocco

Abstract

The nuclear inositol lipid cycle is a well known process, and nuclear phosphoinositide-specific phospholipase C β1 (PLCβ1) signalling activity has been extensively studied in the last decades. We now know that nuclear PLCβ1 is a key player in the control of cell cycle progression; in fact it appears to be involved in the cyclin-mediated regulation of the physiological machinery. Indeed, the recent discovery of a possible involvement of the interstitial deletion of PLCβ1 gene in the progression of myelodysplastic syndrome (MDS) to acute myeloid leukemia in humans (AML) strengthens this contention.

Albeit several papers have reported the techniques used for the study of inositide-dependent signaling in the nucleus, we describe here step by step protocols, which can be followed for the preparation of highly purified nuclei and the subsequent analysis of nuclear PLCβ1 signaling. The described techniques range from nuclear purification to enzymatic activity and to molecular biology methods.

Key words: Phospholipase C β1, Nuclear signaling, PLC activity assay, Nuclei purification, Transmission electron microscopy, PLCβ1a and 1b alternative splicing

1. Introduction

A nuclear signaling, based on lipid hydrolysis, exists and is not a mere duplication of that at the plasma membrane. Among the enzymes of the phosphatidylinositol (PtdIns) metabolism, nuclear phosphoinositide-specific PLCs have been analyzed quite extensively. PLCβ1 plays a key role as a check point in the G1 phase of the cell cycle. Its activation and/or up-regulation is controlled by type 1 insulin-like growth factor receptor (IGF-1R) in both mouse fibroblasts and myoblasts, suggesting that PLCβ1 signaling activity is important for the normal behavior of the cell. The recent discovery of a possible involvement of the interstitial deletion of PLCβ1 gene in the progression of MDS to AML

Christopher J. Barker (ed.), *Inositol Phosphates and Lipids: Methods and Protocols*, Methods in Molecular Biology, vol. 645, DOI 10.1007/978-1-60327-175-2_10, © Humana press, a part of Springer Science+Business Media, LLC 2010

in humans strengthens the contention that nuclear PLCβ1 signaling is essential for physiological cell growth and differentiation. The clinical evolution and the progression of the disease of the MDS patients with PLCβ1 deletion have been worse than expected. In fact, these patients progressed faster than expected to AML although they had a normal karyotype and were considered to have at good prognosis. Therefore, the genetic anomaly affecting a key signaling mediator, i.e., PLCβ1, capable of controlling the cyclin D3/cdk4 checkpoint in G1 phase, seems to be important in the pathophysiology of MDS. The analysis of the role of nuclear PLCβ1 has been furthered by comparing the effect of its overexpression with that of PLCβ1 M2b cytoplasmatic mutant, which is exclusively located in the cytoplasm, in murine erythroleukemia cells. Moreover, in these cells, CD24 gene, coding for an antigen involved in differentiation and hematopoiesis, was up-regulated by nuclear PLCβ1 and not by the cytoplasmic mutant M2b. When PLCβ1 expression was silenced by means of siRNA, CD24 was down-regulated. Up-regulation of CD24 occurred at the transcriptional level, since PLCβ1 affected the CD24 promoter activity. Altogether our findings strengthen the contention that nuclear PLCβ1 constitutes a key step in both cell growth and differentiation.

But it is important to keep in mind that other PLCs can reside, even momentarily, in the nucleus. For example, PLCδ1, which shuttles from nucleus to cytoplasm, could be also involved in the cyclin-dependent machinery, and PLCγ1 acts as a guanine exchange factor in the nucleus activating the nuclear PI3Kinase (1–5).

In this chapter, we are going to outline the methodologies that can be used for the study of nuclear inositol cycle and especially PLCβ1. The most important step is the preparation of highly purified nuclei. In this contribution, we shall describe the two basic systems for obtaining nuclei that we use from cells growing in suspension or in adherence. Thereafter, the other paragraphs describe techniques for measuring PLC activity and for analyzing the expression of PLCβ1 both at the protein and RNA level. At last but not least, we shall also outline the ultrastructural analysis of nuclei by means of transmission electron microcopy.

2. Materials

2.1. Preparation of Nuclei

1. Nuclear isolation buffer: 10 mM Tris–HCl pH7.4; 1% NP-40, 10 mM β-mercaptoethanol; 0.5 mM PMSF; protease inhibitors (complete mini protease inhibitor cocktail Roche, 15 μg/mL calpain inhibitor I, and 7.5 μg/mL calpain inhibitor II).

2. Lysis buffer: buffer (a) 5 mM Tris–HCl pH8.0; 0.1 mM EDTA; 0.06% sodium deoxycholate; 50 mM NaF;10 mM β-Mercaptoethanol; 0.5 mM PMSF; protease inhibitors (mini complete protease inhibitor cocktail Roche, 15 μg/mL calpain inhibitor I, and 7.5 μg/mL calpain inhibitor II).

3. Lysis buffer: buffer (b) M-PER Mammalian protein extraction reagent (PIERCE, catalogue number: 78501); protease inhibitors (as described above). Instead of M-PER it is also possible to use RIPA buffer (plus inhibitors).

4. TM-2 buffer: 10 mM Tris–HCl pH7.4; 2 mM $MgCl_2$; 0.5 mM PMSF; protease inhibitors (as described above) see Note 1.

5. 10% Triton X-100.

6. 1 M $MgCl_2$.

2.2. SDS-Polyacrylamide Gel Electrophoresis (SDS-PAGE) and Western Blot

1. 1.5 M Tris (pH8.8); 0.5 M Tris (pH6.8); 10% SDS solution.

2. 10% APS (Ammonium Persulfate) prepare a 10% solution in water and keep at +4°C for 2 weeks maximum.

3. 30% acrylamide/bis solution (29:1 with 3,3% C) (Bio-Rad). It is a neurotoxin so care should be taken.

4. N,N,N,N′- Tetramethyl-ethylenediamine (TEMED, Bio-Rad).

5. Prestained molecular marker (Bio-Rad). Make 5 μl aliquots and store at –20°C.

6. Running buffer (10×): Prepare 10× stock solution with 144 g glycine, 30 g Tris, and 10 g SDS in 1 L water and store at room temperature. Prepare working solution (1×) by diluting 100 mL stock solution with 900 mL double-distilled water (ddH_2O).

7. Transfer buffer (10×): Prepare 10× stock solution with 144 g glycine, 30 g Tris in 1 L water and store at room temperature. Prepare working solution (1×) by dilution of 100 mL stock with 700 mL water and 200 mL pure methanol (20% v/v). Keep this solution at +4°C (it should be cold when used), and reuse it several times (as long as voltage is maintained constant for each successive run).

8. PBS-T: Prepare Phosphate Buffered Saline 1× with 0.1% (v/v) Tween-20.

9. PVDF membrane (GE-Healthcare) 0.45 μm pore size and 3 MM chromatography paper (Whatman).

10. Optional: Ponceau S (ready-to-use solution from Sigma-Aldrich) and Tris-base 1 M (no need to adjust pH).

11. Blocking solution: 3% (w/v) non fat dry milk, 2% (w/v) BSA (bovine serum albumine) (i.e., Sigma-Aldrich) in PBS-T (see Note 3).

12. Primary antibody: antiPLCβ1 (Santa Cruz catalogue n.sc-9050) diluted 1:750 in PBS-T (see Note 4).

13. Secondary antibody: antirabbit IgG conjugated to horse radish peroxidase (PIERCE).

14. Enhanced chemiluminescent (ECL) reagents (i.e., PIERCE).

15. LaemmLi sample buffer (2× or 4×).

2.3. PLC Activity Assay

1. 800 mM MES buffer pH6.7 (store at –20°C); 800 μM $CaCl_2$ (store at 4°C); PIP_2 1 μg/μl (store at –20°C); (^{3}H)PtdInsP_2 1 μCi/100 μl (store at –20°C); 2% sodium deoxycholate; 1.8% NaCl; 0.6 N HCl (see Note 8).

2. Stop reaction solution is made with chloroform:methanol:HCl (200:100:0.75) (v/v/v) (store at 4°C) (see Note 9).

3. Wash solution: chloroform:methanol:HCl (3:48:47) (v/v/v) (store at 4°C).

4. Solution of chloroform:methanol: H_2O (75:25:2) (v/v/v) (store at 4°C).

5. TLC plate; Whatman 3MM filter paper.

6. Solution 1% of potassium oxalate in methanol-water 1:1, store at 4°C.

7. Scintillation fluid (Ready safe liquid scintillation cocktail for aqueous sample, Beckman).

8. Double-distilled water (ddH_2O).

9. Chloroform:methanol:ammonia solution 30%:H_2O (45:35:2:8) (v/v/v/v) (store at 4°C).

10. Partisil 10 SAX column.

11. 10 mM ammonium phosphate; 1.7 M ammonium phosphate; for both solutions adjust pH to 3.7 with phosphoric acid; they are very expensive, prepare the desired amount for the experiment. Filter before use.

12. (^{3}H)InsP_2 and (^{3}H)InsP_3 (100,000 dpm), as reference compounds, commercially available.

13. EN^{3}HANCE spray (PerkinElmer, catalogue: 6NE97OC).

2.4. PLCβ 1a and 1b Polymerase Chain Reaction (PCR)

1. Sterile double-distilled water (ddH_2O).

2. Template cDNA.

3. PCR amplification kit. There is no requirement for a specific commercial kit; it usually includes 10× $MgCl_2$ free PCR amplification buffer, 25 mM $MgCl_2$, 25 mM dNTP mixture, and *Taq* DNA polymerase.

4. Oligonucleotide primers: F1 forward: 5′-cgggagcacattaaattg-3′; R1 reverse: 5′-gccatgaaagggaaggtt-3′.

5. Agarose for DNA electrophoresis.

6. TAE solution: prepare a 50× TAE stock solution by dissolving 242 g Tris, 57.1 mL glacial acetic acid, 37.2 g Na$_2$EDTA.2H$_2$O in sterile water up to 1 L. Autoclave the stock solution and store at 4°C. Prepare working solution (1×) by diluting 20 mL stock solution with 980 mL distilled water.

7. Ethidium bromide 10 mg/mL, keep this solution in the dark at room temperature. Ethidium bromide is mutagenic, possibly carcinogenic and teratogen, use gloves and special care while handling it or anything that has been in contact with it.

8. Gel loading buffer: it is commercially available (i.e., Promega), no special requirements (to prepare a 6× solution, dilute 0.25% bromophenol blue, 0.25% xylene cyanol FF, 15% Ficoll Type 4000, 120 mM EDTA in sterile water, and keep at 4°C).

9. DNA molecular marker (100 bp).

10. Optional: instead of ethidium bromide it is possible to use SYBER green DNA gel stain (Molecular Probes).

11. Optional: Polaroid black and white instant package films.

2.5. Transmission Electron Microscopy

1. 0.2 M phosphate buffer: mix 28 mL of 0.2 M sodium phosphate monobasic solution with 72 mL of 0.2 M sodium phosphate bibasic solution. Store at 4°C for no longer than 2 weeks (see Note 22).

2. 0.15 M phosphate buffer: dilute 0.2 M phosphate buffer with water 3:1 (v/v). Store at 4°C for no longer than 2 weeks.

3. 50% glutaraldehyde in water commercial stock solution (Fluka, Sigma-Aldrich).

4. 8% paraformaldehyde in water: dissolve 8 g of paraformaldehyde (Sigma-Aldrich) in 100 mL of water and warm it to 60°C under agitation, in a thermal bath for 20 min. The solution will be white and to clarify it add some drops of 1 M NaOH. Let the temperature decrease to room temperature, check the pH (it should be around 7.2 or 7.4), and filter with a 0.2 μm filter (see Note 23). Store at 4°C.

5. Prepare a solution of 4% paraformaldehyde/0.1% glutaraldehyde in 0.1 M phosphate buffer starting from the stock solutions prepared at steps 3 and 4 (see Note 24).

6. Prepare increasing ethanol series solutions (50, 70, and 95%) in distilled water starting from absolute ethanol (Sigma-Aldrich).

7. L.R. White resin is supplied as a single mixture with the catalyst (TAAB, London Resin Co.), store at 4°C.

8. Prepare a solution of 3% uranyl acetate in water. Uranyl acetate takes a long time to dissolve in water. Agitate the solution overnight at room temperature. Store at room temperature and keep away from light. Do not store the solution more than few days, otherwise white crystals will form at the bottom of the bottle (see Note 25).

9. Lead citrate solution: Dissolve 1.33 g of lead nitrate and 1.76 g of sodium citrate in 30 mL distilled water. Agitate for 30 min, and then add 5 mL of 1 N NaOH. The solution will become clear. Stir for 10 min more and add 15 mL of distilled water. Store the solution into a 50 mL sterile syringe, with a filter (0.2 μm) and a needle (25 G or bigger), at 4°C no more than 6 months.

10. 0.5 M Tris base stock solution. Store at 4°C.

11. 0.05 M TBS buffer (0.5 M Tris, 9% NaCl, 1% BSA) pH7.6. Store at 4°C.

12. Blocking solution: 2.5% dry milk (Carnation natural nonfat dry milk, Carnation Company) diluted in 0.05 M TBS buffer.

13. Rabbit antibody antiPLCβ1 (cat: sc-9050; Santa Cruz biotechnology).

14. 0.2 M Tris base stock solution. Store at 4°C.

15. 0.02 M TBS buffer (0.2 M Tris, 9% NaCl, 0.1% BSA) pH 8.2. Store at 4°C.

16. Antimouse IgG 15 nm gold conjugated (British, Biocell International).

3. Methods

3.1. Preparation of Nuclei

Two alternative methods are proposed for the preparation of highly purified nuclear fractions. Depending on the cell line, one method will be more indicated than the other. We use method I for adherent cells such as C2C12 myoblasts, Swiss and NIH 3T3 fibroblasts, 3T3-L1 cells, MCF-7, Saos-2 (1, 6), and method II for suspension cells such as Friend erythroleukemia cell line (7).

3.1.1. Trypsinization of Adherent Cells

1. Examine the flask of cultured cells under an inverted microscope to check for cell viability, contamination, and cell density.

2. Remove medium.

3. Wash the cell monolayer with PBS 1× (1 or 2 times). Let the PBS sit on the cells for at least 30 s in order to remove as much serum as possible.

4. Remove PBS and add a minimal volume (i.e., 0.9 mL for a T-75 flask) of 0.25% w/v trypsin solution, which has been prewarmed to 37°C.

5. Let sit in incubator for about 3 min, and then check the flask under the microscope to make sure that cells have detached from the bottom.

6. When a single cell suspension is obtained, stop trypsin activity by adding about ten times volume of complete medium (with serum).

7. Transfer the cell suspension to a sterile 15 mL conical centrifuge tube.

8. Spin the cells down at 1,200×*g* for 10 min in a tabletop centrifuge.

9. Discard the supernatant carefully; do not disturb the cell pellet.

3.1.2. Nuclei Isolation: Method I

1. Resuspend the cell pellet in 400 μl of ice-cold nuclear isolation buffer, and incubate on ice for 8 min.

2. Add 400 μl of ice cold double-distilled H_2O, and allow cells to swell for 3 min. It is possible to decrease the incubation time if the nuclei look too swollen under the microscope.

3. Shear cells by three passages through a 22-gauge needle and transfer the solution to a 1.5 mL microcentrifuge tube. Check the nuclear preparation on a phase contrast microscope, to make sure that nuclei have lost their outer envelope. Repeat as many passages through the needle as needed, until nuclei are stripped of the envelope, but still intact (see Note 2).

4. Centrifuge the tube at 800×*g* for 10 min at 4°C to recover nuclear fraction: the pellet must result translucid, glassy, and semitransparent.

5. Discard the supernatant, carefully suspend nuclei in 400 μl of ice cold TM2 buffer, and transfer to a new microcentrifuge tube (this step is necessary to wash nuclei from cytoplasmic contamination).

6. Centrifuge the tube at 800×*g* for 10 min at 4°C. The nuclei pellet should look white.

7. Discard as much supernatant as possible: at this point, you have highly purified nuclei.

8. If desired, it is possible to resuspend nuclear preparation with lysis buffer to extract nuclear proteins as described in 3.1.4.

3.1.3. Nuclei Isolation: Method II

1. Examine the flask of cultured cells under an inverted microscope to check for cell viability contamination, cell density, etc.

2. Collect the cells by centrifugation at $1,500 \times g$ for 5 min, wash in PBS 1×, and pellet the cells by centrifugation at $1,500 \times g$ for 5 min.

3. Resuspend the cell pellet with 1 mL of buffer TM-2 (buffer TM-2 10°C).

4. Incubate the cell suspension at room temperature for 1 min and then for 5 min at 0°C (in a container with water and ice).

5. Add 30 µl of 10% Triton X-100 and shear cells by one passage through a 22-gauge needle (see Note 2).

6. Add 5.15 µl of 1 M $MgCl_2$ solution (Mg^{2+} concentration must be brought to 5 mM).

7. Centrifuge the tube at $800 \times g$ for 10 min at 4°C. The pellet should result white.

8. Discard as much supernatant as possible; at this step, you have highly purified nuclei.

9. If desired, it is possible to resuspend nuclear preparation with lysis buffer to extract nuclear proteins as described in subheading 3.1.4.

3.1.4. Nuclear Proteins Extraction

1. Resuspend nuclei in 100 µl of ice cold lysis buffer (you can choose lysis buffer a or b) and transfer the preparation to a new microcentrifuge tube.

2. Lyse nuclei by 30 vigorous passages through a 25-gauge needle keeping the sample always on ice. In the meantime, check nuclear preparation on a phase contrast microscope in order to control whether nuclei are completely lysed: if they are still intact repeat the passages through the needle until nuclei are totally disrupted. It is also possible to lyse nuclei by a quick sonication on ice.

3. Centrifuge nuclei at maximum speed ($\geq 10,000$ unit -g) in a microcentrifuge for 10 min at 4°C; recover the supernatant (nuclear extract) and place it in a new tube.

4. Expected total yield: About 1 µg/µl of nuclear proteins each 10^7 cells (100 µg).

3.2. Electrophoresis and Immunoblotting

3.2.1. SDS-PAGE

1. These guidelines are specific for the use of a minigel system, but may be adapted to a bigger apparatus. Assemble the glass plates with 1.5 mm thick spacer, following the manufacturer's instruction. Optional: pour in water to ensure that the glasses are mounted in a correct (not leaky) way. Before pouring the gel, remove water carefully with a paper towel. Mark on the glass the level that the separating gel should reach, which is the height of the bottom of the wells of the comb plus 1 cm.

Table 1
Solutions for 6% SDS-Polyacrylamide gel electrophoresis

	6% Running gel	6% Running gel	Stacking gel	Stacking gel
Total volume	10 mL (1gel)	15 mL (2 gels)	4 mL (1 gel)	6 mL (2 gels)
H$_2$O	5.3	8.0	2.3	3.37
30% Acrylamide mix	2.0	3.0	0.67	1.0
1.5 M Tris pH 8.8	2.5	3.8	0	0
0.5 M Tris pH6.8	0	0	1.0	1.5
10% SDS	0.1	0.15	0.04	0.06
10% APS	0.1	0.15	0.04	0.06
TEMED	0.008	0.012	0.004	0.006

2. Prepare 10 mL of a 6% acrylamide separating buffer by adding 5.3 mL water to 2.0 mL 30% acrylamide mix, 1.3 mL 1.5 M Tris pH8.8, 0.1 mL 10% SDS, 0.1 mL 10% APS, and 0.008 mL TEMED. Immediately mix the solution and pour it between the glass plates until the mark. On the top of the gel, add 0.2 mL water to exclude oxygen that inhibits the polymerization reaction. Keep the remaining solution to see when the polymerization reaction has occurred (approximately 20 min) (if you need to run two gels simultaneously, refer to Table 1) (see Note 5).

3. Pour off the water, and dry with a paper towel. Prepare 4 mL of stacking gel by adding 2.3 mL water with 0.67 mL 30% acrylamide mix, 1.0 mL 0.5 M Tris pH6.8, 0.04 mL 10% SDS, 0.04 mL 10% APS, and 0.004 mL TEMED. Immediately mix the solution and pour between the glass plates and insert the comb. Avoid air bubbles. Keep the remaining solution in the tube to see when the polymerization reaction has occurred (approximately 20 min).

4. Prepare the samples: To 50 μg of protein (see Note 6), add the right volume of 2× or 4× sample loading buffer. Keep the lysates always on ice.

5. When the stacking gel is polymerized, assemble the gel in the electrophoresis apparatus. Add the running buffer 1× to the chambers. Remove the comb, and rinse the wells with the running buffer, using a syringe or a pipette.

6. Heat the samples and the molecular weight marker at 95°C for 5–10 min, to denature the proteins. Spin samples down and keep on ice while waiting to load them.

7. Load the samples and the molecular weight marker in a defined order.

8. Complete the assembly of the apparatus and connect to a power supply. Run the gel at about 8 V/cm until the blue front has reached the beginning of the separating gel, then increase the speed to 15 V/cm. Run the gel until the blue front has reached the bottom of the separating gel (see Note 7).

3.2.2. Western Blot

1. Proteins separated by SDS-PAGE need to be transferred to a solid support, such as nylon or nitrocellulose membrane, to allow the following detection steps. This transfer occurs electrophoretically, so a blotting apparatus (i.e., Bio-Rad) for minigels is needed.

2. While wearing gloves cut two sheets of 3MM paper the same size of the fiber pads and a sheet of PVDF membrane just larger than the size of the separating gel. Prewet the PVDF membrane with methanol for 5 min and then rinse with transfer buffer 1×.

3. Prepare a tray where the blotting sandwich can be assembled. It is important that every layer of the sandwich is kept wet with the transfer buffer 1× and that air bubbles are avoided, using a glass pipette as a roller or a finger (always wear gloves, use forceps, and handle membranes by their edges).

4. The blotting sandwich must be assembled in the following order: Gel holder cassette (anode side), porous fiber pad, 3MM paper, gel, PVDF membrane, 3MM paper, porous fiber pad, and gel holder cassette (cathode side).

5. In the tray, lay the anode gel holder cassette, the fiber pad, and the 3MM paper.

6. Disconnect the gel unit from the power supply. Discard the stacking gel. Place the running gel on the 3MM paper. Make sure to remember the orientation of the gel in order to trace back the sample loading order.

7. On the top of the gel lay the PVDF membrane, the 3MM paper, the fiber pad, and the cathode gel holder cassette. Close the cassette tightly.

8. Place the cassette in the blotting tank, insert the ice cooling unit, and fill the tank with the transfer buffer 1×. Transfer can occur at 100 mV for 1 h or at 20 mV overnight. It is important to prevent overheating during transfer, so put the tank in an ice bucket or in the cold room.

9. When the blotting is finished, take the cassette out, and disassemble it from the top downward, peeling off each layer. Discard the 3MM papers and the gel. Cut off, with a razor blade, a corner of the PVDF membrane (i.e., bottom right)

in order to remember the sample order. Prestained molecular weight marker should be visible on the membrane.

10. Optional: In order to confirm that proteins have transferred correctly and to help cutting of single lanes, stain the membrane with the Ponceau S for 5 min. Ponceau S can be washed away with Tris-base 1 M and then water, until the membrane returns white.

11. Incubate the PVDF membrane with 10 mL (or a sufficient volume to fully cover the membrane) of blocking solution for 1 h at room temperature with gentle agitation on a platform shaker.

12. Discard the blocking solution and add the primary antibody solution. The antiPLCβ1 antibody (see Subheading 2.2, item 12) can be diluted 1:750 in 10 mL PBS-T. Incubate with a gentle agitation overnight in the cold room (suggested for best signal/background ratio) or at room temperature for about 3 h.

13. Remove the primary antibody (see Note 6). Wash the membrane three times for 5 min with 10 mL PBS-T.

14. Add the fresh secondary antibody solution. The secondary antibody (see Subheading 2.2, item 13) can be diluted 1:10,000 in 10 mL PBS-T. Incubate for 1 h at room temperature with a gentle agitation.

15. Discard the secondary antibody. Wash the membrane three times for 15 min with 10 mL PBS-T.

16. Place the membrane on a flat plastic surface. Mix the ECL reagents, and add the solution to the membrane. Incubate for 5 min. Ensure that during incubation, the ECL solution covers completely and evenly on the membrane.

17. Place the membrane in an X-ray cassette and expose the membrane to an X-ray film or a PhosphoImager station (example of PLCβ 1a and 1b western blot is showed in Fig. 2a).

For ref. see (2, 8).

3.3. Enzymatic Activity and Chromatography

3.3.1. PLC Activity Assay

1. This assay is set-up for 60 μg of nuclear proteins in a final volume of 200 μl. The reaction starts when protein is added to the complete reaction mix. Therefore, add proteins as last step.

2. Perform the PLC assay in triplicate. Prepare a number of glass vials corresponding to samples.

3. Prepare the PLC assay buffer B: Mix 25 μl of 800 mM MES buffer pH 6.7 to 25 μl of 800 μM $CaCl_2$. At the end, you will have a 50 μl mix of 400 mM MES buffer pH 6.7 and of 400 μM $CaCl_2$, which is needed for one sample. Prepare enough sample buffer B for the number of samples +2.

4. Add to each vial 50 µl of PLC assay buffer B. Leave vials on ice.

5. Add ddH_2O to each vial to reach a final volume of 200 µl (keep in mind that proteins should be added at the end).

6. Substrate mix C is composed of $PtdInsP_2$ (^3H) $PtdInsP_2$, sodium deoxycholate, and NaCl (steps 6–9). Prepare in a glass vial the substrate for the PLC assay, which is a mix of $PtdInsP_2$ and (^3H) $PtdInsP_2$; for each sample, mix 100.000 dpm of (^3H) $PtdInsP_2$ with 3 nmol of $PtdInsP_2$. Remember that usually 1 µ Ci = 2.200.000 dpm in 100 µl. Therefore, for one sample use a mix of 4.5 µl of (^3H) $PtdInsP_2$ (1 µCi/100 µl) and 3 µl of $PtdInsP_2$ 1 µg/µl.

7. Mix and evaporate to dryness with a stream of nitrogen (see Note 11).

8. After evaporation the residue is dissolved in 6 µl of 2% sodium deoxycholate. Mix well by vortexing for at least 2 min.

9. Add 66.6 µl of 1.8% NaCl and mix well again by vortexing for at least 2 min. Sonicate the substrate mix (C) in a bath sonicator for 5 min. Alternatively, the lipid sample can be vortexed vigorously for 4 min. At the end, the opaque solution must become transparent. This procedure yields a suspension of multilamellar vesicles.

10. Add 72.6 µl of solution C to each sample glass vial and then add the nuclear proteins. Note: the sensitivity of the assay depends not only on the turnover rate of the enzyme but also on the duration of the assay, which must be kept exactly the same for each sample. Make sure to allow the same reaction time for each sample. The reaction is carried out at 37°C for 30 min (preferably on a thermomixer). Prepare enough substrate mix C for the number of samples +2.

	Blank (input)	Sample
Amount of protein	0	60 µg (i.e.,40 µl)
B (assay buffer)	50 µl	50 µl
C (substrate mix)	72,6 µl	72,6 µl
H_2O	77.4 µl	37.4 µl
Total reaction volume	200 µl	200 µl

11. Stop the reaction by adding of 1 mL of stop solution. Vortex each tube for 5 s. When all reactions are stopped, add 300 µl of 0.6 N HCl, vortex, and centrifuge at room temperature for 5 min at 2,500 × g.

12. At the end of the centrifugation, remove carefully an equal amount of the aqueous top phase from each sample, without disturbing the organic phase (i.e., 100 μl).

13. A small aliquot (20 μl) of the upper phase, (containing inositol polyphosphates), can be collected directly in scintillation vials containing 5 mL emulsifier scintillation fluid for radioactivity measurements. Blanks containing no proteins are included in each series to correct for nonenzymatic hydrolysis. Liberated inositol phosphates, mainly InsP_3, are calculated as nmol per μg of the protein used as enzyme source.

14. The remaining 80 μl of the upper phase is centrifuged for 5 min at high speed ($\geq$10,000×g), to clarify it. The radiolabeled water-soluble metabolites are separated using an HPLC system equipped with a radioactivity flow detector. Supernatant is injected into a prepacked Partisil 10× SAX column at room temperature. The column is eluted with a linear gradient from distilled water to 1.7 ammonium phosphate at 1.5 mL/min. The labelled inositols are automatically displayed by scintillator fluid FloOne (Pakard). The HPLC profile could show some InsP_2 coming from degradation of the substrate (see Fig. 1a).

15. Wash the organic phase twice with 1 mL of wash solution, vortex, centrifuge, and remove the upper phase.

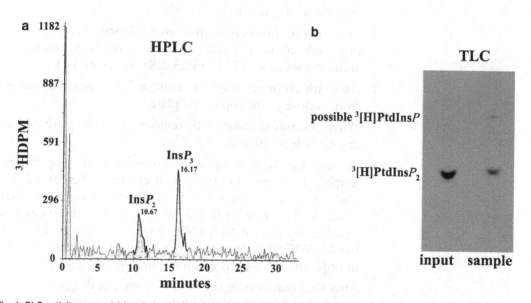

Fig. 1. PLC activity assay: (a) inositol polyphosphates, generated by PLC activity assay can be separated by HPLC. (b) TLC fluorography shows the starting amount of ^{3}(H) PtdInsP_2 (input) present in the assay mix and the ^{3}(H) PtdInsP_2 hydrolyzed by PLC activity by the sample.

16. Vacuum dry the organic phase mixture in a Speedvac evaporator for approximately 40 min.

17. Resuspend the dried mixture in 50 µl of chloroform:methanol:H_2O (75:25:2) solution.

18. The phospholipids in the organic phase can be analyzed by thin-layer chromatography (TLC) on plates coated with silica gel. The PtdInsP_2, which has not been hydrolyzed, can be separated by TLC and its amount can be determined by radioactivity measurements. At this point, it is also possible to freeze samples at –20°C.

For ref. see (1, 7).

3.3.2. Thin-Layer Chromatography (TLC)

1. Impregnate (activate) the TLC plate with a solution 1% of potassium oxalate in methanol-water 1:1. Activation involves driving off water molecules that bond to the polar sites on the plate.

2. Incubate the activated TLC plate in an oven for 3 h at 120°C.

3. Equilibrate the TLC tank: Add 190 mL chloroform, 10 mL methanol, and a piece of Whitman 3MM filter paper approximately the size of the TLC sheet to the tank and let stand for 2 h.

4. Spot the samples, 5 µl at a time, 2 cm above the edge of a plastic-backed TLC sheet and 2 cm one from the other (see Notes 10 and 11).

5. Develop the chromatogram in an equilibrated chromatography tank containing 200 mL of chloroform: methanol: ammonia solution: H_2O (45:35:2:8) (see Note 12).

6. Allow the chromatography to run for 2 h or until the solvent front is close to the top of the plate.

7. When the run is completed, remove the TLC plate and air dry for at least 30 min.

8. To visualize the TLC plate, it is possible to develop fluorographically. Spray TLC plate with enhancer (be careful it is toxic, spray it in a fume hood). Allow the plate to dry, then cover the TLC plate with plastic wrap in an autoradiography cassette, and place it on it a film. Expose the film at –70°C up to 48 h. While exposing the film make a mark (i.e., right bottom) in order to recognize the film orientation (see Fig. 1b).

9. After the fluorography, the regions corresponding to PtdInsP_2 are scraped from the plates and counted by scintillation. Add scrapings into prefilled scintillation vials; vortex vials and count spots 24 h later, in order to stabilize (reduce) luminescent background signals (luminex) (see Note 13).

For ref. see (9).

3.4. PCR Analysis

3.4.1. PLCβ 1a and 1b Polymerase Chain Reaction (PCR)

1. PLCβ 1a and 1b differ in their 3′ spliced sequence: exon 32 is present in 1b isoform but not in the 1a isoform, whereas exon 33 is present in 1a isoform but not in the 1b isoform. Nucleotide sequence can be found in Pub Med, using the following accession number for the CoreNucleodide records:

Human PLCβ 1a	AJ278313
Human PLCβ 1b	AJ278314
Mouse PLCβ 1a	U85712
Mouse PLCβ 1b	U85714
Rat PLCβ 1a	M20636
Rat PLCβ 1b	L14323

A first screening to detect the possible presence of PLCβ 1a and 1b can be made by PCR using primers that anneal at the 3′ region, in order to obtain 2 bands that differ in size, so that can be distinguished by electrophoresis (see Note 14).

2. Thaw all the components on ice; *Taq* polymerase is stored in glycerol, so it should be taken from −20°C just when it needs to be added to the mix. In a tube prepare a master mix for $n + 1$ samples (where n is the number of samples) containing all the components except for the cDNA: water, forward primer, reverse primer, dNTP, 10× buffer, $MgCl_2$ and *Taq* polymerase (see Notes 15, 16 and 18). Aliquot the master mix in the n single thin-walled PCR tubes.

Component	Final concentration
10× buffer	1×
F1 primer	200 nM
R1 primer	200 nM
dNTP mix	0.2 mM
Taq polymerase	1.25 U
$MgCl_2$	2 mM
ddH_2O	Up to 30 μl

3. Add cDNA to the sample tubes. Quickly mix and spin down at 4°C (see Note 17).

4. Place tubes in a thermal cycler. Set the annealing temperature, extension time, and cycle number:

Denature cDNA at 95°C for 2 min then repeat these three amplification steps for 35 cycles: perform a final extension step at 72°C for 5 min.

- denaturation: 95°C for 30 s
- annealing: 43°C for 30 s
- extension: 72°C for 60 s

5. At the end of the PCR collect the tubes from the thermal cycler and analyze the products by electrophoresis (PCR products can also be stored at 4°C for few days).

3.4.2. PCR Gel Electrophoresis

1. Assemble the gel tray and the comb (comb should remain 2 cm from the edge of the tray and its wells 2 mm above the surface of the tray). If no tray holder is available, wrap the tray with tape.

2. Prepare a 1% agarose gel by dissolving 1 g of agarose in 100 mL (volume may vary depending on the tray size) of TAE 1× into a glass beaker. To dissolve it, stir on a hot plate or use microwave.

3. Wait until the gel temperature is decreased (when is possible to handle the beaker without getting burnt), then add ethidium bromide to a final concentration of 0.5 µg/mL. Swirl the solution until the ethidium is uniformly mixed. Pour the warm gel solution in the tray and wait until the gel is solidified, approximately 20 min (see Notes 19 and 20).

4. Place the tray in the horizontal electrophoresis apparatus, remove the comb (and the tape), and fill the chamber with TAE 1× until the buffer just covers the agarose gel.

5. Add loading buffer to the samples. It is possible to add loading buffer directly to the PCR tube or to prepare a parafilm strip, add drops of loading buffer (for the *n* samples), mix the PCR products with the loading buffer on parafilm, and directly load the samples into wells.

6. Load the samples and the molecular weight marker into wells.

7. Run the gel at 100 mV; after approximately 45 min check the run.

8. Visualize the PCR products on a UV transilluminator (see Note 21). Photograph the PCR bands with a Polaroid or digital camera (example of PLCβ 1a and1b PCR is shown in Fig. 2b).

3.5. Electron Microscopy

3.5.1. Sample Preparation for Transmission Electron Microscopy

1. This protocol is optimized for Friend erythroleukemia cells (10), but can be adapted for other cell lines. Cells cultured in 75-cm² tissue flasks with GI medium are collected and centrifuged at $1,000 \times g$ for 8 min. Resuspend cell pellets in PBS in order to obtain a cell concentration of 1×10^6 cells/mL. Transfer 500 µl of cell suspension in a 500 µl centrifuge tube

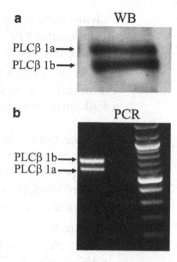

a   WB

PLCβ 1a⟶
PLCβ 1b⟶

b   PCR

PLCβ 1b⟶
PLCβ 1a⟶

Fig. 2. PLCβ 1a and 1b expression in C2C12 cells. (**a**) Nuclear lysate was resolved by 6% acrylamide gel (SDS-PAGE) and then analyzed by western blot with antibody antiPLCβ1. (**b**) cDNA was amplified by PCR using primers able to distinguish both isoforms.

(see Note 26). Centrifuge the cells at $1,000 \times g$ for 8 min to obtain cell pellets.

2. With a glass Pasteur pipette remove the PBS solution gently and add 500 µl of fixative solution 4% paraformaldehyde/0.1% glutaraldehyde in 0.1 M phosphate buffer (see Note 27). Add the fixative solution very slowly avoiding the resuspension of pellet and let the samples at 4°C for 1 h (see Note 28).

3. From now on, never resuspend cell pellets in the solutions that you will use in the following steps. Following three brief washes in 0.15 M phosphate buffer, each of 500 µl, at room temperature, cell pellets are dehydrated by increasing ethanol series. Incubate cell pellets with ethanol solutions, from 50 to 70% and 95%, twice for 10 min for each solution, at 4°C.

4. Wash three times for 10 min each with 100% ethanol at 4°C (see Note 29). After the third wash in absolute ethanol, detach the fixed pellet from the inner side of the microcentrifuge tube with the help of a tooth stick. This will help the following shifting in gelatin capsule described at step 6.

5. Remove ethanol (leaving the pellet in the bottom) and add 500 µl of a solution of L.R. White resin and 100% ethanol 1:1 previously prepared. Leave the samples at room temperature for 1 h and then substitute the solution with a fresh one made of L.R. White resin and 100% ethanol 3:1, overnight at 4°C.

6. The following morning change the solution twice with pure L.R. White, incubate each time for 3 h, and place cell pellets in gelatin capsules (see Note 30).

7. Polymerization is performed at 40°C in an oven for 3 days. Attention must be taken to ensure that the cell pellets are not exposed to temperature more than 55°C to preserve antigenicity.

8. L.R. White embedded samples are sectioned with an ultramicrotome (for example, Reichert JUNG FC 4/E; Leica). Sections of 90 nm are mounted on nickel grids and air dried for 1 day.

3.5.2. Immunogold Labeling for PLCβ1

9. Rinse sections, three times for 5 min, by placing grids on 100 μl droplets of 0.05 M TBS, pH7.6. After these washing steps, perform a blocking step by incubating the grids on drops of blocking solution (see Note 31) for 30 min at room temperature (see Note 32).

10. Incubate the sections on droplets containing the antiPLCβ1 primary antibody diluted 1:20 in blocking reagent, overnight at 4°C (see Note 33). The incubation must be performed in a humid chamber (lay droplets of the primary antibody on a piece of Parafilm, then place over a wet filter paper inside a petri dish).

11. The following morning rinse sections three times for 5 min with large droplets of 0.05 M TBS buffer, then once with 0.02 M TBS buffer, pH8.2 (see Note 34). Incubate sections on droplets of the secondary antibody antirabbit IgG 15 nm gold conjugated, diluted 1:10 in 0.02 M TBS buffer pH 8.2, at room temperature for 1 h and 30 min. Rinse the specimens on large droplets of 0.02 M TBS buffer, three times for 5 min, followed by washes with distilled water, three times for 5 min at room temperature.

12. Stain the sections with droplets of 3% uranyl acetate, in the dark at room temperature for 10 min. Avoid air drying. Rinse sections with water and stain again with lead citrate solution, in the dark at the room temperature for 10 min. After one wash in distilled water, let the sections air dry at least for 30 min.

13. A strong electron beam can destroy the L.R. resin during the observation of the sections. Before observing, coat the sections with a thin layer of carbon, using a metal evaporation device (BAL-TEC AG.).

14. Observe sections with a transmission electron microscopy (TEM), we use a Philips CM10 (FEI Company).

4. Notes

1. Each cell line has its own distinctive features; therefore, it is necessary to adapt these protocols. An important detail to keep in mind is the presence of $MgCl_2$. The damage of nuclear envelope is prevented by the presence of magnesium. In a cell

line, if nuclei are particularly breakable it is advisable to increase the final concentration of magnesium.

2. These protocols can be adapted to various cell lines. During the nuclear preparation constantly check, on a contrast phase microscope, how the extraction is going. If necessary, change the incubation time in each step.

3. The blocking solution is optimized for antiPLCβ1 Santa Cruz antibody; different antibodies may require different blocking solutions such as 5% nonfat dry milk or 3% BSA.

4. This rabbit antibody is suitable to detect both isoforms PLCβ 1a and 1b. Other antibodies are available from different commercial suppliers. Once used, the primary antibody can be saved for other western blots by addition of 0.05% (v/v) sodium azide and storage at 4°C. Sodium azide prevents bacterial growth; make a 5% sodium azide solution in PBS (keep at 4°C, in the dark) that is a 100× stock solution. As a rule, we usually keep the used antibodies for as long as they still work.

5. A 6% acrylamide gel is recommended to separate the 2 isoforms: PLCβ 1a is 150 kD whereas 1b is 140 kD.

6. The amount of loaded protein depends on the abundance of PLCβ1 in the studied system, so the right quantity should be optimized for each cell line. PLCβ1 is not usually abundant, so it is advisable to load as much as possible. It is also possible to enrich PLCβ1 content by purification of nuclei (see Subheading 3.1).

7. To separate the 2 isoforms it is possible to run the gel for a longer time, but avoid running the gel too long, otherwise the proteins will leave the gel. You may control the presence of high molecular weight proteins by checking the presence of the corresponding bands in the prestained molecular marker.

8. There are common rules while working with radioactivity samples. Wear gloves, lab coat, goggles, and radioactivity dosimeters when handling radioactive samples. Discard all materials that have come in direct contact with radioactivity in the radioactive waste. Radioactivity experiments should be performed in a specialized area, when possible.

9. Be careful when working with chloroform; chloroform evaporates easily into the air; always work under a chemical fume hood.

10. While spotting the samples try to keep the sample dot as small as possible. Spot the samples, 5 μl at a time, in order to give the solvent time to evaporate and not to form a big dot.

11. Choose the proper volume of chloroform: methanol: ammonia solution: H_2O keeping in mind that it is very important that the spots are above the level of solvent.

12. Start drying keeping the stream of nitrogen far enough away to avoid spilling of radioactive solution.

13. During PLC activity assay it is possible that substrate (^{3}(H) PIP$_2$) degradation occurs. This degradation leads to unspecific ^{3}(H)PIP formation (see Fig. 1b).

14. Remember that protein PLCβ 1a is bigger (150 kD) than PLCβ 1b (140 kD), while PCR product PLCβ 1a is shorter than PLCβ 1b (length depends on species). The following protocol is based on rat sequence, but can be adapted for human or mouse. In this case, PLCβ 1a PCR product will be 400 bp while PLCβ 1b will be 540 bp.

15. Care should be taken while working with DNA: Try to avoid DNA cross-contamination and dNTPs degradation by working always on ice, wearing gloves, using tips with aerosol filters, and autoclaving all the tubes.

16. MgCl$_2$ concentration can influence notably PCR amplification; if necessary, optimize MgCl$_2$ concentration testing different concentration, i.e., 1.5, 3.0, or 4.5 mM.

17. cDNA concentration depends on the abundance of PLCβ1 in the studied system. A too low starting cDNA concentration can result in a failed amplification, but a too high starting cDNA concentration can result in amplification of nonspecific PCR products, and possibly a Taq polymerase inhibition. We usually reverse transcribe 1 μg of mRNA in 30 μl of mix; from this cDNA we usually collect 1 μl that we amplified by PCR. For genomic DNA, it is possible to start from 0.1 to 1 μg. If the template cDNA is not highly purified, you may try to use more Taq polymerase, up to 3 U (be careful, as higher Taq polymerase concentrations can result in amplification of nonspecific PCR products).

18. It is possible to increase the primer concentration (up to 300 nM) but primer-primer dimer formation can occur when small amounts of template are used; in these cases, PCR efficiency will decrease.

19. If the gel solution is too hot, ethidium bromide can evaporate, so prevent personal proximity with ethidium steam. If possible, place the gel tray under a fume hood.

20. Instead of ethidium bromide it is possible to use SYBER safe DNA gel stain (the solution is a 10,000× stock solution). SYBER is less carcinogenic than ethidium, but it may be (there are different solutions available) less sensitive than ethidium in visualizing PCR products. If using a digital system, check also presence of the right filter for SYBER green.

21. Ethidium bromide present in the gel allows fluorescent visualization of the PCR products under UV light.

22. Phosphate buffers allow the growth of bacteria and molds. Always check the solutions before use.

23. Do not store stock solution of 8% paraformaldehyde in water more than 1 week at 4°C. Paraformaldehyde molecules tend to polymerize with time.

24. Always use a fresh made fixative solution of 4% paraformaldehyde/0.1% glutaraldehyde, which should be prepared just before the experiment.

25. Caution: the uranyl acetate is based on uranyl compounds, which have high levels of the radioactive isotopes! Always wear gloves and work under a chemical fume hood.

26. 1.5 mL centrifuge tubes can be used to prepare cell pellets. The centrifuge tubes should be filled completely.

27. The fixative solution of 4% paraformaldehyde/0.1% glutaraldehyde is recommended for the preservation of antigenicity. Do not increase the percentage of glutaraldehyde. Glutaraldehyde has a greater power to cross link proteins and undoubtedly stabilizes tissue structure more strongly than paraformaldehyde, but it also induces a reduction in antigenicity.

28. Maintaining the temperature of 4°C is very important during the fixation process to reduce autolysis processes in the cell pellets. Autolysis processes are generally faster at higher temperature such as room temperature or 37°C.

29. Do not leave cell pellets in ethanol solutions longer than necessary. Ethanol could extract cell material and damage cell structure.

30. Shift cell pellet from microcentrifuge tubes to gelatin capsules with a toothpick or a glass Pasteur pipette, which has previously been broken in the middle. Do not worry if you break the pellet during the procedure. Even small pieces of cell pellets could be enough material for transmission electron microscopy analysis.

31. The best conditions for the blocking step should be determinated depending on the experimental system. The blocking solution can be changed by increasing or decreasing the percentage of single components such as dry milk and BSA, depending on the background signal of the samples.

32. All the immunogold labeling steps are performed by placing grids on droplets of the different solutions. Spot the droplets on Parafilm and work on it for the washing and for the blocking steps. The suggested volume for the droplets is 100 μl, but it can be increased to perform better washes.

33. Optimal antibody concentration should be determined by the user. With Friend erythroleukemia cell line and the antiPLCβ1 antibody of Santa Cruz Biotechnology Inc. (cat: # sc-9050),

we found that the optimal dilution (in which background and unspecific signal are minimal) is 1:20.

34. Check carefully the pH of the 0.02 M TBS buffer, it should be pH 8.2. Gold conjugated antibodies are more stable in solution with an alkaline pH.

Acknowledgments

The authors own work was supported by grants from Italian MIUR-FIRB (HumanProteome Net), MIUR-COFIN, and the Carisbo Foundation.

References

1. Martelli, A. M., Gilmour, R. S., Bertagnolo, V., Neri, L. M., Manzoli, L., and Cocco, L. (1992) Nuclear localization and signalling activity of phosphoinositidase C beta in Swiss 3T3 cells. *Nature* 358, 242–245.

2. Faenza, I., Matteucci, A., Manzoli, L., Billi, A. M., Aluigi, M., Peruzzi, D., Vitale, M., Castorina, S., Suh, P. G., and Cocco, L. (2000) A role for nuclear phospholipase Cbeta 1 in cell cycle control. *J Biol Chem* 275, 30520–30524.

3. Fiume, R., Faenza, I., Matteucci, A., Astolfi, A., Vitale, M., Martelli, A. M., and Cocco, L. (2005) Nuclear phospholipase C beta1 (PLCbeta1) affects CD24 expression in murine erythroleukemia cells. *J Biol Chem* 280, 24221–24226.

4. Follo, M. Y., Finelli, C., Clissa, C., Mongiorgi, S., Bosi, C., Martinelli, G., Baccarani, M., Manzoli, L., Martelli, A. M., and Cocco, L. (2009) Phosphoinositide-phospholipase C beta1 mono-allelic deletion is associated with myelodysplastic syndromes evolution into acute myeloid leukemia. *J Clin Oncol* 27, 782–790.

5. Suh, P. G., Park, J. I., Manzoli, L., Cocco, L., Peak, J. C., Katan, M., Fukami, K., Kataoka, T., Yun, S., and Ryu, S. H. (2008) Multiple roles of phosphoinositide-specific phospholipase C isozymes. *BMB Rep* 41, 415–434.

6. O'Carroll, S. J., Mitchell, M. D., Faenza, I., Cocco, L., and Gilmour, R. S. (2009) Nuclear PLCbeta1 is required for 3T3-L1 adipocyte differentiation and regulates expression of the cyclin D3-cdk4 complex. *Cell Signal* 21, 926–935.

7. Martelli, A. M., Billi, A. M., Gilmour, R. S., Neri, L. M., Manzoli, L., Ognibene, A., and Cocco, L. (1994) Phosphoinositide signaling in nuclei of Friend cells: phospholipase C beta down-regulation is related to cell differentiation. *Cancer Res* 54, 2536–2540.

8. Faenza, I., Bavelloni, A., Fiume, R., Lattanzi, G., Maraldi, N. M., Gilmour, R. S., Martelli, A. M., Suh, P. G., Billi, A. M., and Cocco, L. (2003) Up-regulation of nuclear PLCbeta1 in myogenic differentiation. *J Cell Physiol* 195, 446–452.

9. Cocco, L., Gilmour, R. S., Ognibene, A., Letcher, A. J., Manzoli, F. A., and Irvine, R. F. (1987) Synthesis of polyphosphoinositides in nuclei of Friend cells. Evidence for polyphosphoinositide metabolism inside the nucleus which changes with cell differentiation. *Biochem J* 248, 765–770.

10. Fiume, R., Ramazzotti, G., Teti, G., Chiarini, F., Faenza, I., Mazzotti, G., Billi, A. M., and Cocco, L. (2009) Involvement of nuclear PLCbeta1 in lamin B1 phosphorylation and G2/M cell cycle progression. *FASEB J* 23, 957–966.

Chapter 11

Methods to Assess Changes in the Pattern of Nuclear Phosphoinositides

Nullin Divecha

Abstract

Phosphatidylinositol (PtdIns) and its phosphorylated derivatives represent less than 5% of total membrane phospholipids in cells. Despite their low abundance, they form a dynamic signalling system that is regulated in response to a variety of extra and intra-cellular cues (Curr Opin Genet Dev 14:196–202, 2004). Phosphoinositides and the enzymes that synthesize them are found in many different sub-cellular compartments including the nuclear matrix, heterochromatin, and sites of active RNA splicing, suggesting that phosphoinositides may regulate specific functions within the nuclear compartment (Nat Rev Mol Cell Biol 4:349–360, 2003; Curr Top Microbiol Immunol 282:177–206, 2004; Cell Mol Life Sci 61:1143–1156, 2004). The existence of distinct sub-cellular pools has led to the challenging task of understanding how the different pools are regulated and how changes in the mass of lipids within the nucleus can modulate nuclear specific pathways. Here we describe methods to determine how enzymatic activities that modulate nuclear phosphoinositides are changed in response to extracellular stimuli.

Key words: Nuclear isolation, In nuclei labeling, Phosphatidylinositol-5-phosphate, Phosphatidylinositol-4-phosphate, Phosphoinositides, Nuclear lipid signalling

1. Introduction

Understanding how nuclear phosphoinositides are regulated is crucial to understanding their role in regulating nuclear specific processes such as transcription, chromatin remodeling, and mRNA splicing and export (1–4). These nuclear pathways are intimately associated with a number of human pathologies, such as the development of cancer, diabetes and mental illness; and the kinases that modulate nuclear phosphoinositides are therefore potential targets for the development of novel therapeutics. Phosphoinositide kinases, like all other kinases, are very

Christopher J. Barker (ed.), *Inositol Phosphates and Lipids: Methods and Protocols*, Methods in Molecular Biology, vol. 645,
DOI 10.1007/978-1-60327-175-2_11, © Humana press, a part of Springer Science+Business Media, LLC 2010

"druggable" but have the added benefit that their ATP binding sites differ significantly from protein kinases, thus enabling the development of more specific inhibitors. Thus far, there are no nuclear specific phosphoinositide kinase isoforms, suggesting that they shuttle between the cytosol and the nucleus and therefore unravelling how they are regulated specifically in the nucleus will yield the most potential for the development of novel drugs. As yet, there are no methods to specifically study the activity of nuclear enzymes in vivo without the requirement for the isolation of nuclei. This in itself triggers the first problem: that of obtaining clean, plasma membrane free, nuclei. In general, suspension cells such as murine erythroleukamia (MEL) cells yield extremely clean nuclei compared to nuclei isolated from adherent cells such as fibroblasts, which require the use of stringent detergents that are likely to also remove enzymes and lipids from within the nucleus. Once the nuclear fraction has been isolated, enzymatic activities can be assayed in different ways. The enzymes can be extracted and assayed using exogenously added substrate. An alternative and highly successful method has been to use the nuclei intact and allow endogenous enzymes to phosphorylate endogenous lipids (in nuclei labeling) (5–8). This assay is carried out in the presence of labeled ATP, after which the labeled lipids are isolated and can be analyzed in a variety of ways. Although the interpretation of the results is complicated, and is discussed later, the beauty of this assay is that it can maximize the possibility of discovering changes in nuclear phosphoinositides in response to various environmental cues. Here we detail the use of such an assay to investigate whether there are changes in nuclear lipid metabolism in response to differentiation, stimulation with extracellular stressors, or as a consequence of progression through the cell cycle.

2. Materials

2.1. Cell Culture and Nuclear Isolation

1. Dulbeco's modified Eagles Medium (DMEM) (Invitrogen) supplemented with 10% fetal bovine serum.
2. Nocodazole (1 mg/mL stock solution in DMSO stored at –20°C).
3. Swell buffer (5 mM Tris–HCl pH 7.4, 1.5 mM KCl, 2.5 mM $MgCl_2$). This can be made as a 20 times stock and stored at 4°C.
4. 1.8 M sucrose (molecular biology grade and RNAse and DNAse free).
5. 1 M $MgCl_2$.
6. 33 mM EGTA.

7. Cushion buffer (10 mM Tris–HCl pH 7.4, 1 mM EGTA, 1.5 mM KCl, 5 mM $MgCl_2$, 460 mM sucrose). This can be made as a 20 times stock without the sucrose and stored at 4°C.

8. Final resuspension buffer (FRB 10 mM Tris–HCl pH 7.4, 1 mM EGTA, 1.5 mM KCl, 5 mM $MgCl_2$, 290 mM sucrose). This can be made as a 20 times stock without the sucrose and stored at 4°C.

9. Nuclear membrane strip solution (10 mM Tris–HCl pH 7.4, 1 mM EGTA, 1.5 mM KCl, 5 mM $MgCl_2$ 290 mM sucrose, 0.3% Triton X-100).

10. Swing out centrifuge (Hereus megafuge 1.0) cooled to 4°C.

11. 10 mL Polyallomer centrifuge tubes (Beckman 326814).

2.2. In Nuclei Labeling of Phosphoinositides

1. Phosphatidylserine (PtdSer 10 mM in chloroform. Sigma p3660), Phosphatidylinositol (PtdIns Sichem or Echelon), phosphatidylinositol-4-phosphate (PtdIns4P Sichem or Echelon), and phosphatidylinositol-5-phosphate (PtdIns5P Sichem or Echelon). Synthetic commercial phosphoinositides are normally produced as ammonium salts, which are relatively insoluble in chloroform. In order to solubilize the lipids we routinely convert them to the acidic form (see Note 1).

2. ATP (stock 5 mM dissolved in 10 mM Tris pH 7.4 10 mM $MgCl_2$ and stored in aliquots at −20°C).

3. ^{32}P-ATP specific activity of at least 3,000 Ci/mmol. Appropriate precautions must be taken when handling (^{32}P) e.g., gloves, Perspex screens, and personal radiation monitoring.

4. Stop solution: 50 mL chloroform, 50 mL methanol, 400 µl of Folch extract (Sigma type I Folch fraction from bovine brain (030K7063) at a stock concentration of 1 mg/mL in chloroform). The stop solution can be stored at −20°C in a glass bottle.

5. TLC plates: Silica gel TLC plates (Merck 1.05721.0001) are dipped in a solution of 1% potassium oxalate, 1 mM EDTA, and then placed in an oven (110°C) for 1 h (see Note 2).

6. TLC developing solution (chloroform:methanol:28% ammonia: H_2O; 45:35:2:8) should be prepared fresh, placed into a TLC tank containing two pieces of Whatman 3MM paper (20×20 cm), and allowed to equilibrate for 1 h before plate development.

2.3. Determination of the Isomer of PtdInsP Generated During in Nuclei Labeling

1. PtdIns4P and PtdIns5P lipid standards (Sichem or Echelon) adjusted to 1 µmol/mL.

2. Phosphatidylserine (10 mM in chloroform. Sigma p3660).

3. Phosphatidic acid (3 mM in chloroform. Sigma).

4. 2× PIPkin buffer: 100 mM Tris–HCl pH 7.4, 20 mM MgCl$_2$, 2 mM EGTA, 140 mM KCl.

5. GST-PIP4Kalpha (phosphatidylinositol-5-phosphate 4-kinase alpha) purified after expression in bacteria (see Note 3).

6. GST-mPIP5Kalpha (murine phosphatidylinositol-4-phosphate 5-kinase alpha) purified after expression in HEK293 mammalian cells (see Note 3).

7. Stop solution: 50 mL chloroform, 50 mL methanol, 400 µl of Folch extract (Sigma type I Folch fraction from bovine brain (030K7063) at a stock concentration of 1 mg/mL in chloroform). The stop solution can be stored at –20°C in a glass bottle.

2.4. Mass Assay of Phospholipid Phosphate

1. Clean high temperature-resistant glass tubes.

2. Concentrated perchloric acid.

3. 2.5% ammonium molybdate.

4. 10% ascorbic acid freshly prepared in water (0.5 g in 5 mL).

3. Methods

All organic solvents should be of the highest quality available (HPLC grade or better), and high purity water (Milli-Q) should be used for the preparation of all aqueous solutions. All operations are performed at room temperature unless otherwise stated.

3.1. Growth, Differentiation, and Stress Stimulation of MEL Cells

1. MEL cells are routinely maintained at a density of 0.2–1.0×10^6 cells/mL in DMEM-10% FBS at 37°C.

2. MEL cell differentiation: MEL cells are diluted to 0.1×10^6 cells/mL and then either maintained as control cells or differentiated by the addition of 1.5% DMSO. After 4 days, 90% of the cells are differentiated as assessed by staining for hemoglobin.

3. UV irradiation or Cisplatin stimulation of MEL cells: MEL cells are diluted to 0.2×10^6 cells/mL on day 1, and on day 2 are exposed to UV irradiation or treated with cisplatin. For UV irradiation, cells (100 mL) are centrifuged ($352 \times g$) and the cell pellet resuspended in 5 mL of DMEM-FBS and spread on a 10 cm diameter culture plate. The lid is removed and the cells are irradiated with 500 J/m^2 using a Stratagene DNA cross linker. The cells are diluted to 100 mL and maintained in culture for 1 h (9).

4. Cell cycle progression. Cells are diluted to 0.2×10^6 cells/mL (2×10^7 cells are required per time point), and the next evening nocodazole (100 ng/mL) is added and left for a

maximum of 10 h (see Note 4). The cells are collected by centrifugation and washed once with warm PBS. The cells are finally resuspended in DMEM-FCS at a density of approximately 0.5×10^6/mL. MEL cells progress through the cell cycle in a synchronous manner and at various time points, cells are collected for nuclear isolation. It takes approximately 4 h for the cells to pass from the nocodazole block to S-phase and 9 h to reach G2/M. 1 mL of cells should be collected for the determination of the position of the cells in the cell cycle using FACS analysis (cells are collected by centrifugation, fixed in 70% ethanol (−20°C), stained with propidium iodide, and treated with RNAase).

3.2. Nuclear Isolation

1. 2×10^7 cells are pelleted by centrifugation ($352 \times g$) and washed once with ice cold PBS (25 mL). The PBS is decanted and the remaining PBS is removed using a tissue.

2. The cells are resuspended in 5 mL of swell buffer and maintained on ice for 5 min. The cells are then disrupted by two passages through a 22 gauge needle using a 6 mL syringe before being transferred to a 10 mL polyallomer centrifuge tube (see Note 5).

3. To stabilize the nuclei 800 µl of 1.8 M sucrose, 150 µl of 33 mM EGTA, and 25 µl of 1 M MgCl$_2$ are added, and the nuclei are pelleted by centrifugation (4 min, $352 \times g$, 4°C) (see Note 6).

4. The supernatant is removed (see Note 7) and the nuclei are resuspended in 1 mL of ice cold FRB and 5 mL of ice-cold cushion is layered directly underneath. The nuclei are pelleted through the sucrose cushion by centrifugation (4 min, $352 \times g$, 4°C) and washed once with 5 mL of ice cold FRB.

5. To remove the nuclear membrane the nuclear pellet is resuspended in 5 mLs of nuclear membrane strip solution and incubated on ice for 5 min. The nuclei are pelleted by centrifugation, washed once with 5 mL of ice cold FRB, and finally resuspended in 300 µl of ice cold FRB.

6. The protein concentration is determined using a Bradford assay (Biorad) and compared to BSA standards.

3.3. In Nuclei Labeling of Phosphoinositides

1. 5–20 µg of nuclei (either with or without nuclear membranes) are resuspended in 50 µl of FRB, and incubated at 30°C for 2 min. The reaction is started by the addition of 50 µl of FRB containing 100 µM ATP and 10 µCi ^{32}P-ATP. The reaction is allowed to proceed for 2–10 min (see Note 8).

2. The reactions are stopped by the addition of 1 mL of stop solution followed by 100 µl of H$_2$O and 250 µl 2.4 M HCl. The tubes are vigorously shaken and centrifuged, and the lower phase, containing the radiolabelled phosphoinositides,

is removed into a new Eppendorf tube and dried by rotary evaporation. The upper phase that contains the majority of the free [^{32}P]-ATP is discarded.

3. [^{32}P]-labeled phosphoinositides and phosphatidic acid are separated by TLC. 0.5 cm pencil lines, 1 cm apart, are marked 1.5 cm above the bottom of the TLC plate. The samples are resuspended in 8 µl of chloroform:methanol (90:10) and spotted onto the pencil marks. The Eppendorf tube is washed with a further 8 µl of chloroform:methanol (90:10), and the wash is spotted on top of the original sample. The spotted samples are allowed to dry for 10 min at room temperature before development in a TLC tank containing 70 mLs of TLC developing solution until the solvent front reaches 4 cm from the top.

4. Quantitation of the radioactivity in [^{32}P]-labelled lipids is carried out using a phosphoimager (we routinely use the Fuji BAS reader) or by exposing the dried TLC to X-ray film followed by scraping off the silica containing the [^{32}P]PtdIns$(4,5)P_2$ and quantitation by Cerenekov counting. Densitometry of the exposed developed X-ray film is not recommended.

Differentiation, treatment with cisplatin or progression through the cell cycle induces changes in the in nuclei labeling patterns of phosphoinositides and phosphatidic acid (Fig. 1). For example, upon differentiation, there is an increase in the labeling in PtdInsP_2. Deacylation (10) of the labeled PtdInsP_2 and analysis of the released headgroup by HPLC or by separation on PEI cellulose plates demonstrated that PtdIns$(4,5)P_2$ rather than PtdIns$(3,4)P_2$ or PtdIns$(3,5)P_2$ is the only detectable radiolabeled isomer generated in MEL cell nuclei. Treatment with cisplatin and progression through the cell cycle induce increases in both the labeling of PtdInsP_2 and PtdIns$P$. Changes in the labeling of PtdIns$(4,5)P_2$ or PtdIns$P$ could be due to changes in the levels of the substrate, subnuclear localization of the kinase, activity of the enzyme itself, or in the activity of another enzyme that is responsible for the removal of the product. To begin to define what exactly may be responsible for these changes, we add back exogenous substrate and re-assay the nuclear activity (described below). Further analysis would include mass measurements of relevant nuclear substrates (9) and the solubilization and purification of the enzymatic activity either using specific antibodies or conventional chromatography.

3.4. Changes in Enzymatic Activity

To determine whether changes in substrate levels or enzymatic activity are responsible for changes in nuclear phosphoinositide labeling, high concentrations of exogenous substrates are added to purified nuclei and the in nuclei labeling is re-assessed. For example, addition of exogenous PtdIns4P to isolated nuclei led to the same amount of synthesis of labeled PtdIns$(4,5)P_2$ in control nuclei as compared to nuclei from differentiated cells (5).

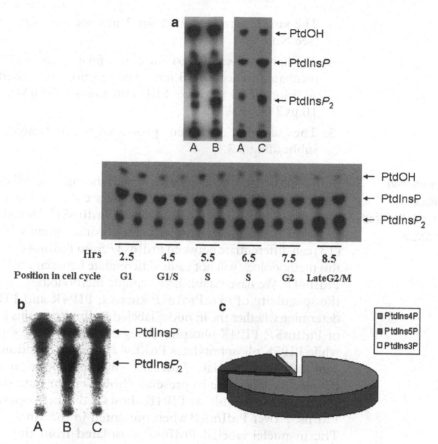

Fig. 1. (**a**) *Top panel.* Nuclei were isolated from control cells (A) or from DMSO differentiated cells (B) or after treatment with cisplatin (C). The nuclei were labeled with ^{32}P-ATP, after which the radiolabelled lipids were separated by TLC. The identity of the various lipids is shown. *Bottom panel.* MEL cells were arrested in the cell cycle using nocodazole for 10 h after which the cells were washed and released into DMEM-FCS. At the times indicated nuclei were isolated. When all of the nuclear samples had been collected they were labeled with ^{32}P-ATP and the radiolabelled lipids were separated by TLC. The data show that there is an increase in PtdIns$P$ and PtdInsP_2 synthesis as cell progress into S-phase and again at the point of G2M. (**b**) Radiolabelled PtdIns$P$ was isolated after separation by TLC and then incubated alone (*lane 1*) with PIP5K (*lane 2*) or PIP4K (*lane 3*) together with cold ATP. The lipids were re-isolated and separated by TLC. The data demonstrate that incubation with either PIP4K or PIP5K leads to the synthesis of PtdInsP_2 demonstrating that the initial labeled PtdIns$P$ contained both radiolabelled PtdIns$4P$ and PtdIns$5P$. Separation by PEI cellulose also demonstrated that a small proportion, approximately 4% of the labeled PtdIns$P$ contained PtdIns$3P$. The data were quantitated and are shown as a pi-chart (Fig. 1**c**).

This suggested that upon differentiation there was an increase in the mass of nuclear PtdIns$4P$. Indeed further studies using specific assays to determine the mass of PtdIns$4P$ demonstrated that there was a threefold increase in the levels of nuclear PtdIns$4P$ upon differentiation (11, 12).

1. For each sample, PtdSerine (10 nmol), PtdIns (5 nmole), PtdIns$4P$ (1 nmol), and Diacylglycerol (3 nmol) are mixed and dried by rotary evaporation.

2. The lipids are resuspended with 40 µl of 10 mM Tris pH 8.0 and left at room temperature for 15 min to rehydrate.

3. The samples are sonicated for 2 min using a bath sonicator (see Note 9).

4. The lipid mix is added to nuclei to a final volume of 50 μl and incubated on ice for 15 min. The reaction is initiated by the addition of 50 μl of 2× FRB containing 100 μM ATP and 10 μCi of ^{32}P-ATP.

5. The samples are then processed as indicated in the subheading 3.3.

3.5. Determination of the Isomer of PtdInsP Synthesized During in Nuclei Labeling

Both cisplatin treatment and progression through the cell cycle lead to increased labeling of PtdInsP. There are three known isomers of PtdInsP, PtdIns3P, PtdIns4P, and PtdIns5P. Deacylation of the labeled PtdInsP and subsequent chromatography by HPLC (13) can differentiate between PtdIns3P from PtdIns4P. However, this methodology will not easily differentiate between PtdIns5P and PtdIns4P. We have established a simple methodology, which uses the specificity of two PtdInsP-kinases, PIP4K and PIP5K, to determine whether the in nuclei labeled PtdInsP contains PtdIns4P or PtdIns5P. PIP4K phosphorylates PtdIns5P on the 4-position, while PIP5K phosphorylates PtdIns4P on the 5-position. PIP4K can also phosphorylate PtdIns3P on the 4-position and can be used to determine its presence; however, the data should be interpreted cautiously as PIP4K shows a distinct preference for PtdIns5P over PtdIns3P when presented in the same lipid mix. The in nuclei labeled PtdInsP is isolated from the silica gel TLC plate and then phosphorylated using either PIP4K or PIP5K in the presence of nonlabeled ATP. Generation of labeled PtdIns(4,5)P_2 determines the initial presence of labeled PtdIns4P and PtdIns5P in the original mix.

1. Directly after phosphorimaging the PtdInsP spot is located and the silica gel is scraped into a clean eppendorf using a sterile scalpel blade (see Note 10).

2. 0.95 mLs of chloroform:methanol:2.4 M HCL (0.25:0.5:0.2) is added, and the solution is sonicated in a water bath for 2 min. The mixture is then left at room temperature for 10 min.

3. The silica is removed by centrifugation (1 min full speed in a microfuge), and the supernatant is collected in a new eppendorf. 0.25 mLs of chloroform and 0.25 mLs of water are added, the mixture is shaken, and the two phases are separated by centrifugation (1 min full speed in a microfuge). The lower phase, containing the labeled lipid is removed to a clean eppendorf and dried by rotary evacuation.

4. The dried labeled PtdInsP is resuspended in 500 μl of chloroform, and 50 μl is removed into a new Eppendorf tube for PtdIns4P analysis. The remaining 450 μl are used to measure PtdIns5P.

5. 10 nmoles of phosphatidylserine (1 μl) and 3 nmoles of phosphatidic acid (1 μl) are added to every tube, and the lipids are dried by rotary evaporation.

6. 50 μl of diethylether is added to each tube followed by 100 μl of 10 mM Tris–HCl pH 7.4. The Eppendorf tubes are then capped, vortexed, and sonicated in a bath sonicator (20 s each tube). The lipid solution is then gently centrifuged (300 rpm, $8 \times g$) in an Eppendorf microfuge for 30 s) before the diethyl-ether is removed by rotary evaporation for 5 min.

7. A master-mix is prepared containing 2×PIPkin, 100 μM ATP, 0.5 μg GST-PIP4K (for the PtdIns5P assay), or GST-PIP5K (for the PtdIns4P assay). 100 μl of the master mix solution is added to each tube to start the reaction and incubated at 30°C for 2 h.

8. The reactions are stopped by the addition of 1 mL of stop solution followed by 250 μl 2.4 M HCl. The tubes are vigorously shaken, centrifuged and the lower phase, containing the radiolabelled phosphoinositides, is removed into a new Eppendorf tube, and dried by rotary evaporation.

9. [^{32}P]PtdIns(4,5)P_2 is separated from other phosphoinositides by TLC as described above (see Note 11).

10. Quantitation of the radioactivity in [^{32}P]PtdIns(4,5)P_2 is carried out using a phosphoimager (we routinely use the Fuji BAS reader) or by exposing the dried TLC to X-ray film followed by scraping off the silica containing the [^{32}P] PtdIns(4,5)P_2 and quantitation by Cerenekov counting.

Figure 1b shows that in nuclei labeled PtdIns$P$ comprised of approximately 75% PtdIns4P, 19% PtdIns5P, and 4% PtdIns3P. The level of PtdIns3P should be interpreted cautiously as the PIP4K has a preference for the phosphorylation of PtdIns5P over PtdIns3P when both are present. The simplest interpretation of these data is that nuclei can synthesize PtdIns4P and PtdIns5P through the 4 and 5 phosphorylation of Phosphatidylinositol.

3.6. Mass Assay of Phospholipid Phosphate

1. A 1 mM solution of potassium phosphate is used to generate a standard curve (0, 5, 10, 20, 50, 100, 150 nmoles) of phosphate and together with the samples derived from the supernatants from the neomycin-coated bead column are dried in ultra clean (phosphate-free) 15 mL glass tubes (see Note 12) in an 80°C oven overnight.

2. 50 μl of concentrated perchloric acid is added to each tube followed by incubation at 180°C for 30 min using a heating block in a fume cupboard.

3. After cooling to room temperature, 250 μl of water, 50 μl of 2.5% (w/v) ammonium molybdate, and 50 μl of freshly

prepared 10% (w/v) ascorbic acid are sequentially added and incubated at room temperature until color development is complete (at least 2 h). 150 μl is pipetted in duplicate into the wells of a flat-bottomed transparent 96 well microtitre plate and the absorbance is measured at a wavelength of 630 nm using a spectrophotometer plate reader. Quantitation of phospholipid phosphate in the unknown samples is performed by comparison with the standard curve.

4. Notes

1. To convert phosphoinositides into their acidic form the dried lipid is resuspended in 1 mL of chloroform/methanol (1:1 v/v) transferred into an Eppendorf tube and 200 μl H_2O and 250 μl 2.4 M HCl are added. The tube is vigorously shaken, centrifuged (maximum speed for 1 min in an Eppendorf microfuge), and 750 μl of the upper phase is removed and discarded. The lower phase is washed with 750 μl of TUP. The lower phase is removed into a new Eppendorf tube. 10 μl of lower phase is dried in a clean glass tube and used to measure the total amount of phospholipid phosphate (subheading 3.6). PtdIns4P and PtdIns5P each contain two phosphate groups. Therefore, the number of moles obtained in the phosphate assay is twice the number of moles of starting lipid. For PtdIns$(4,5)P_2$ the amount of phosphate represents three times the number of moles of starting lipid. The remainder of the lower phase is dried and resuspended in chloroform (adjusted to 1 nmole lipid/mL) and stored at –20°C in a clean glass tube closed with a Teflon coated cap (Supelco).

2. Preparation of TLC dipping solution: A 1.5 L solution of 2% potassium oxalate, 2 mM EDTA in water is diluted with 1.5 L of methanol. This solution is then placed in a spare TLC tank, and TLC plates are dipped in one direction (the undipped area is designated on the top of the plate and should be marked for orientation) and dried in an oven (110°C). The dipping solution can be stored in the TLC tank at 4°C for at least 3 months.

3. cDNA clones encoding GST-PIP4K and GST-PIP5K and protocols for their isolation are available on request from N. Divecha.

4. We have found that leaving MEL cells in nocodazole for more than 10–12 h leads to an increase in the proportion of the cells that are irreversibly blocked in G2 and an increase in the number of cells that undergo endoreplication.

5. Nuclei can be isolated from smaller numbers of cells in which case the cells are resuspended in 1 mL of swell buffer and disrupted using a 1 mL syringe. The sucrose cushion buffer

can be layered carefully directly underneath the disrupted cell solution and centrifuged as stated. For different cell types, different gauge needles may yield cleaner nuclei. For example, passage of HeLa cells through a 25 gauge needle yields much purer nuclei than through a 23 gauge needle. The purity of nuclei can be rapidly assessed microscopically. Clean nuclei display a rounded morphology, whereas contaminated nuclei display an abundance of cytoskeletal material attached to the nuclei. A more thorough assessment of contamination can be carried out by Western blotting analysis using antibodies directed against actin (Chemicon, MAB-150-1R) and tubulin (anti-α-tubulin Sigma T5168). However, care should be taken as numerous studies have demonstrated the presence of actin in nuclei and in a number of chromatin remodelling complexes.

6. If nuclei are unstable (manifested by an inability to pellet and to resuspend easily the nuclei during the wash procedures), the addition of spermine (0.15 mM) and spermidine (0.1 mM) immediately after hypotonic lysis should alleviate this problem. We have also found that some batches of sucrose tend to increase nuclear instability, and therefore we always use molecular biology grade sucrose (Sigma) that is both RNAse and DNAse free.

7. The supernatant contains the light (plasma membranes and endoplasmic reticulum) and the heavy (mitochondria, lysosomes and rough endoplasmic reticulum) membranes, which can be further fractionated by sucrose gradient centrifugation, can be isolated by centrifugation at $100,000 \times g$ for 1 h.

8. It is important to demonstrate that the incorporation of radiolabel into nuclear phosphoinositides is linear with respect to the amount of nuclear protein and the time of the assay.

9. Samples can be sonicated directly into 10 mM Tris–HCl pH 7.5 using a bath sonicator. We routinely utilize the Bioruptor sonicator (Diagenode). The sonication is carried out for 2 min at high frequency using a pulse time of 20 s on and 15 s off. It is also possible to solubilize lipids using diethyl ether. In this case, the dried lipid is solubilized using 50 μl of diethyl ether. The required aqueous phase is then added and the solution is vigorously vortexed. The tube is centrifuged briefly, and the ether is removed by rotary evaporation for 5 min.

10. It is essential that the TLC plate is not allowed to completely dry and that elution of the lipid from the silica is carried out as soon as possible. The plate should be wrapped in saran wrap to expose it to the phosphoimager plate after which the silica should be scraped within 2 h.

11. In vitro PIP5K can phosphorylate a number of different phosphoinositides including PtdIns3P. However, using biological samples we have never observed the formation of

$[^{32}P]PtdIns(3,5)P_2$ $(PtdIns(3,5)P_2$ is separated from PtdIns $(4,5)P_2$ using the TLC system described). In vitro PIP4K can phosphorylate both $PtdIns5P$ and $PtdIns3P$; however, it shows a marked preference for $PtdIns5P$. In the TLC system described, $PtdIns(3,4)P_2$ will migrate slightly below $PtdIns(4,5)P_2$. To distinguish better $PtdIns(3,4)P_2$ from PtdIns $(4,5)P_2$, the simplest strategy is to deacylate the lipids and to separate the corresponding glycerophosphoinositol-bisphosphates using PEI cellulose TLC plates (Merck). Briefly, the dried lipids are deacylated by the addition of 200 µl of monomethylamine reagent (100 µl of monomethylamine (fluka 41% in H_2O) and 100 µl of methanol) and incubation with continuous mixing for 50 min at 53°C. The monomethylamine is removed by rotary evaporation (this takes approximately 2–3 h), and the glycerophosphoinositol-bisphosphates are then dissolved in 200 µl of H_2O. This solution also contains the fatty acids, which are removed by washing with 200 µl of a mixture of n-butanol:petroleum ether (bp 40–60):ethylformate (20:4:1, v/v/v). The tubes are shaken vigorously, centrifuged, and the upper phase is discarded. Another 150 µl of the organic solvent mixture is added, mixed thoroughly, and centrifuged, and the lower aqueous phase is removed into a clean 0.5 mL Eppendorf tube and dried by rotary evaporation. The glycerophosphoinositol-bisphosphates are dissolved in 2 µl of 10 mM potassium phosphate and are spotted onto a PEI cellulose TLC plate 1.5 cm from the bottom. The Eppendorf tube is washed with 2 µl of 10 mM potassium phosphate that is then spotted on top of the original spotted sample. The TLC plate is developed in a tank containing 70 mL of 0.48 M HCl. Standards of deacylated $PtdIns(3,4)P_2$ and $PtdIns(4,5)P_2$ are easily generated using GST-PIP4K to phosphorylate 1nmole of either $PtdIns3P$ or $PtdIns5P$ under the conditions outlined in the $PtdIns5P$ assay followed by their deacylation.

12. It is possible to clean tubes after each use; however, to avoid contamination artifacts, we use cheap disposable glass tubes.

References

1. Jones, D.R., and Divecha N. (2004) Linking lipids to chromatin. *Curr. Opin. Genet. Dev.* **14**, 196–202.

2. Irvine, R.F. (2003) Nuclear lipid signalling. *Nat. Rev. Mol. Cell Biol.* **4**, 349–360.

3. Hammond, G., Thomas, C.L., and Schiavo, G. (2004) Nuclear phosphoinositides and their functions. *Curr. Top. Microbiol. Immunol.* **282**, 177–206.

4. Martelli, A.M., Fala, F., Faenza, I., Billi, A.M., Cappellini, A., Manzoli, L., and Cocco, L. (2004) Metabolism and signaling activities of nuclear lipids. *Cell. Mol. Life Sci.* **61**, 1143–1156.

5. Cocco, L., Gilmour, R.S., Ognibene, A., Letcher, A.J., Manzoli, F.A., and Irvine, R.F. (1987) Synthesis of polyphosphoinositides in nuclei of Friend cells. Evidence for

polyphosphoinositide metabolism inside the nucleus which changes with cell differentiation. *Biochem. J.* **248**, 765–770.

6. Cocco, L., Martelli, A.M., Gilmour, R.S., Ognibene, A., Manzoli, F.A., and Irvine RF. (1988) Rapid changes in phospholipid metabolism in the nuclei of Swiss 3T3 cells induced by treatment of the cells with insulin-like growth factor I. *Biochem. Biophys. Res. Commun.* **154**, 1266–1272.

7. Sun, B., Murray, N.R., and Fields, A.P. (1997) A role for nuclear phosphatidylinositol-specific phospholipase C in the G2/M phase transition. *J. Biol. Chem.* **272**, 26313–26317.

8. Vann, L.R., Wooding, F.B., Irvine, R.F., and Divecha, N. (1997) Metabolism and possible compartmentalization of inositol lipids in isolated rat-liver nuclei. *Biochem. J.* **327**, 569–576.

9. Jones, D.R., Bultsma, Y., Keune, W.J., Halstead, J.R., Elouarrat, D., Mohammed, S., Heck, A.J., D'Santos, C.S., and Divecha, N. (2006) Nuclear PtdIns5P as a transducer of stress signaling: an in vivo role for PIP4Kbeta. *Mol. Cell* **23**, 685–695.

10. Clarke, N.G., and Dawson, R.M. (1981) Alkaline O leads to N-transacylation. A new method for the quantitative deacylation of phospholipids. *Biochem. J.* **195**, 301–306.

11. Martelli, A.M., Cataldi, A., Manzoli, L., Billi, A.M., Rubbini, S., Gilmour, R.S., and Cocco L. (1995) Inositides in nuclei of Friend cells: changes of polyphosphoinositide and diacylglycerol levels accompany cell differentiation. *Cell. Signal.* **7**, 53–56.

12. Divecha, N., Letcher, A.J., Banfic, H.H., Rhee, S.G., and Irvine, R.F. (1995) Changes in the components of a nuclear inositide cycle during differentiation in murine erythroleukaemia cells. *Biochem. J.* **312**, 63–67.

13. Stephens, L.R., Hughes, K.T., and Irvine, R.F. (1991) Pathway of phosphatidylinositol(3,4,5)-trisphosphate synthesis in activated neutrophils. *Nature* **351**, 33–39.

Chapter 12

Measurement of Phosphoinositide 3-Kinase Products in Cultured Mammalian Cells by HPLC

Frank T. Cooke

Abstract

The phosphoinositide 3-kinase (PI3K) family catalyses the addition of a phosphate group to the D-3 position of polyphosphoinositides (PPIn). Since the discovery in the late 80s that phosphatidylinositol is phosphorylated in the D-3 position in eukaryotic cells, there has been an explosion of interest in these PPIn. Although the four D-3 PPIn (phosphatidylinositol 3-phophate (PtdIns3P), PtdIns(3,4)P_2, PtdIns(3,5)P_2 and PtdIns(3,4,5)P_3) represent only a small proportion of PPIn, production of D-3 PPIn is required for an ever-increasing number of processes. Measurement of the PPIn levels in intact cells cultured cells has been vital to our understanding of the metabolism and function of these important signalling molecules; methods are described herein that allow measurement of PPIn levels in cultured cells, with emphasis on the 3-OH PPIn.

Key words: PI3K, Phosphoinositides, HPLC, Isotopic labelling

Abbreviations

BSA	bovine serum albumin
cpm	counts per minute
dpm	disintegrations per minute
Etn	ethanolamine
FAF	fatty acid-free
Gro$P$	glycerophospho
HBBSS	HEPES buffered balanced salts solution
HPLC	high performance liquid chromatography
Ins	inositol
Ptd	phosphatidyl
PI3K	phosphoinositide 3-kinase
PPIn	polyphosphoinositide
P_i	inorganic PO_4
TBAS	tetrabutyl ammonium hydrogen sulphate

Christopher J. Barker (ed.), *Inositol Phosphates and Lipids: Methods and Protocols*, Methods in Molecular Biology, vol. 645, DOI 10.1007/978-1-60327-175-2_12, © Humana press, a part of Springer Science+Business Media, LLC 2010

1. Introduction

The 7 phosphorylated derivatives of phosphatidylinositol (PtdIns), often collectively referred to as polyphosphoinositides (PPIn), are a minor component of eukaroytic cell membranes (PPIn metabolism is summarized in Fig. 1); nevertheless, their synthesis is needed for an ever-increasing spectrum of cellular processes including: regulation of the actin cytoskeleton, chemotaxis, protein trafficking, glucose uptake, organelle acidification, and signal transduction (1). Although PtdIns4P and PtdIns(4,5)P_2 were discovered in the 50s (2), catalysed by the advances in high performance liquid chromatography (HPLC) technology, it has been only relatively recently found that PtdIns is also phosphorylated in the D-3 position (3). Subsequently, a family of enzymes, the phosphoinositide 3-kinase family (PI3K), has been characterized that synthesize D-3 PPIn (Table 1).

A vital tool in the study of the metabolism and function of PI3K products has been a reliable method for assaying them in cultured cells. Although nonisotopic labelling based methods have been published for determining PtdIns(3,4,5)P_3 levels (4–6), isotopic labelling and subsequent HPLC analysis are still the method of choice for measuring levels of PPIn *in cells*. While it is likely that advances in other methodologies, such as mass spectrometry, may eventually prevail (7, 8) (see Chapter 13), using the methods described below 3-phosphorylated PPIn have

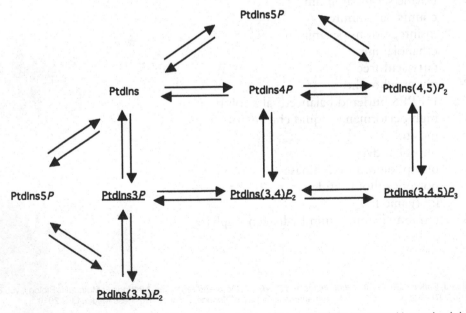

Fig. 1. The complex metabolism of the PPIn. Interconversion between the various PPIn is very rapid, so only a brief period of labelling with [^{32}P]-P$_i$ is necessary to reach equilibrium. Inositol incorporates more slowly into PtdIns, but will label all PPIn to the same extent.

Table 1
The isoforms of the mammalian PI3K family of enzymes. The type I isoforms are acutely responsive to extra-cellular stimuli (see Table 4), and are responsible for synthesis of the second messenger PtdIns(3,4,5)P_3. The type II isoforms have been suggested to synthesize PtdIns3P in vivo, although recent data also imply PtdIns(3,4)P_3 as an in vivo product. The type III PI3K is homologous to the only PI3K in yeast, Vps34. To date it seems that PtdIns(3,5)P_2 is not synthesized directly by a PI3K isoform, but from phosphorylation of PtdIns3P

	Sub classes	In vivo product
Type IA	α	PtdIns (3,4,5)P_3
	β	
	δ	
Type IB	γ	
Type II	C2a	PtdIns3P(?)
	C2β	
Type III	hVps34	PtdIns3P

been measured reproducibly in a wide variety of cell types, for examples, see (9–12).

To measure PPIn, first, cells are isotopically labelled, and this can be achieved with either [^{3}H]-inositol, or [^{32}P] inorganic PO$_4$ ([^{32}P]-P$_i$) – or indeed [^{33}P]-P$_i$ (see Note 1). Ideally, PPIn are labelled to isotopic equilibrium, so that the total label incorporated into each PPIn is proportional to its mass; the pros and cons of equilibrium labelling are discussed in the first overview chapter. (^{3}H)-inositol incorporates specifically into inositol containing lipids making subsequent analysis easier to interpret, and PtdIns will be labelled to the same extent as the other PPIn, which can be useful in final data analysis (Fig. 2). [^{32}P]-labelled D-3 PPIn standards, which are simple and cheap to generate in vitro, can be added to [^{3}H]-labelled samples to facilitate peak identification. Additionally, [^{3}H] produces negligible ionising radiation. However, [^{3}H]-inositol is expensive, and incorporating sufficient label into PPIn to allow detection of 3-OH PPIn can be challenging and dependent on cell type. Isotopic equilibrium typically takes 48 h to achieve, although pilot experiments in your cell line of choice can be undertaken. As many cell types are not tolerant of inositol withdrawal, which is necessary for efficient incorporation of label, labelling to isotopic equilibrium may not always be possible.

[^{32}P]-P$_i$ is relatively cheap and incorporates into PPIn rapidly, with equilibrium typically achieved within 1 h. However, as [^{32}P]-P$_i$ incorporates to some extent into the entire cellular phospholipid complement chromatographic analysis can be harder to interpret, and [^{3}H]-PPIn standards are expensive. Incorporation of [^{32}P]-P$_i$ into PtdIns is poor. Furthermore, there is the additional

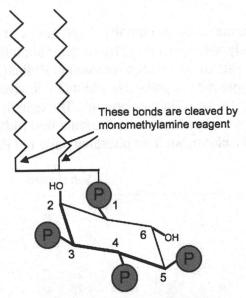

Phosphatidylinositol 3,4,5-trisphosphate

Fig. 2. Schematic diagram of PtdIns(3,4,5)P_2. The phosphates in the 3, 4, and 5 positions will label rapidly; however, the 1 position labels more slowly. [^{32}P]-P_i-labelling will result in a varying specific activity between the PPIn – PtdIns will be poorly labelled whereas PtdIns(3,4,5)P_3 will have approximately three times the label incorporated as PtdInsP_n. The site of cleavage of monomethylamine reagent is also shown.

problem of hazardous ionising radiation. Although both [^{3}H]-inositol and [^{32}P]-P_i labelling have been successfully employed to measure 3-OH PPIn (9, 11, 13), and methods are presented for both, [^{32}P]-P_i labelling is the preferred method of this author.

Once the cells have been labelled and stimulated/manipulated, they are killed and the lipids are recovered using a two-phase organic solvent extraction. Maintaining quantitative recovery in this stage is paramount to the success of the experiment. Due to their charge at neutral pH PtdInsP_2 isomers and PtdIns(3,4,5)P_3 are quite water soluble and thus prone to remaining in the aqueous phase. Thus, three strategies are employed encourage them into the organic phase and maximize their recovery: First, the extraction procedure described by Folch is employed (14). This uses a ratio of 8:4:3 (v/v/v) CHCl$_3$:MeOH:aqueous and was designed specifically for extraction of acidic phospholipids. Many researchers mistakenly use the extraction procedure described by Bligh and Dyer (15), which uses a ratio of 1:1:0.9 (v/v/v) CHCl$_3$:MeOH:aqueous; this gives poor recovery of PtdInsP_2 isomers and PtdIns(3,4,5)P_3 and should not be used. Second, the aqueous phase is acidified to 1 M HCl, promoting protonation of acidic phospholipids thus reducing their charge. Third, tetrabutylammonium hydrogen sulphate (TBAS) is added to the extraction; TBAS is a quaternary lipophilic ammonium salt, and aids neutralization of acidic compounds in the organic phase. In the author's experience, PtdInsP_2 isomers and PtdIns$(3,4,5)P_3$ cannot be extracted efficiently without addition of TBAS. Additionally, 20 µg unlabelled brain phospholipid extract is added to each sample before extraction.

After extraction the PPIns are deacylated to remove the fatty acyl chains and the water-soluble glycerophosphoinositol (Gro$P$Ins) head groups (Fig. 3) are separated by high performance

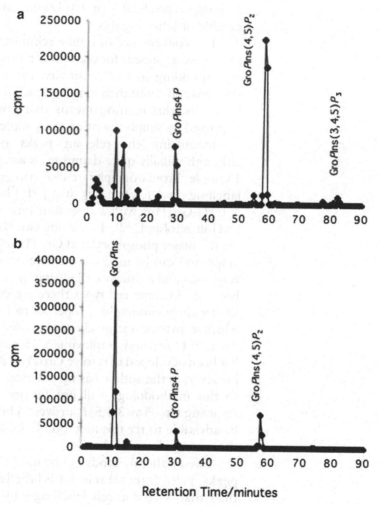

Fig. 3. Typical chromatographs of COS-7 cells labelled with [^{32}P]-P$_i$ and Skov-3 cells labelled with [^{3}H]-inositol. Notice that despite 8 years between performing these experiments that the retention times are very similar. (a) As expected [^{32}P]-P$_i$ labelling results in more peaks on the chromatograph, and PtdIns, which by mass is far more abundant than any other PPIn is poorly labelled. Gro$P$Ins(4,5)P_2 contains the most label, followed by Gro$P$Ins4P. It is best to use these peaks to orientate yourself on the chromatograph. (b) [^{3}H]-labelling produces much fewer peaks, especially at the start of the chromatograph, making orientation easier. Gro$P$Ins is the most prominent peak. In this experiment, insufficient label was incorporated to detect the PtdInsP_2 isomers PtdIns(3,5)P_2 and PtdIns(3,4)P_2.

liquid chromatography (HPLC) on a strong anion exchange (SAX) column. The Gro$P$Ins heads groups are detected by liquid scintillation counting.

The author has obtained acceptable results using a variety of, often venerable, HPLC apparatus, so it is difficult to provide a recommendation in this regard. It should be bourn in mind that the gradient given in this methodology starts with a ramp from 1% to 7% salt over 30 min. With a two-pump HPLC system and flow rate of 1 mL/min, this would require the pumps to be able to pump reproducibly at 10 µl/min. Most modern systems are capable of achieving this.

The convenience of online scintillation detection provides a persuasive argument for use of this method. However, due to the corresponding loss of sensitivity, the author prefers to collect fractions and count them on a static scintillation counter. Although laborious, this method means that counting accuracy can be improved by simply counting the samples for longer.

Identifying the relevant peaks on the chromatograph, although initially quite daunting, is actually easier than it seems. Example chromatographs are shown in Figs. 3 and 4. [^{3}H]-inositol labelling gives fewer peaks than [^{32}P$_i$]-labelling. The major peak is [^{3}H]-Gro$P$Ins with a retention time around 8–12 min. With [^{3}H]-inositol and [^{32}P]-P$_i$ labelling Gro$P$Ins4P and Gro$P$Ins(4,5)P_2 are the major phosphorylated Gro$P$InsP$_n$ peaks on the chromatograph and can be used to orientate oneself. Gro$P$Ins(3,4,5)P_3 is very polar and is usually well separated from any other compounds; however, in some cell types there are two additional peaks with similar chromatographic properties to Gro$P$Ins(3,4,5)P_3, both of which seem to be side-products of the deacylation of PtdIns(4,5)P_2. An HPLC method employing a 25 cm Whatman SAX column has been developed to resolve Gro$P$Ins(3,4,5)P_3 from these peaks; however, as the author has no practical experience of this system so this methodology will not be presented here. However, if resolving Gro$P$Ins(3,4,5)P_3 proves to be a problem then it would be advisable to try this method (12). Sample preparation will be the same as described herein.

Two further methods can be used to identify D-3 Gro$P$InsP$_n$ peaks. First, internal standards labelled with an isotope other than that used for cell labelling can be added to samples – e.g., [^{32}P]-labelled standards can be added to samples labelled with [^{3}H]. The different energy spectrums of [^{3}H] and [^{32}P] allow them to be differentiated in the same sample by liquid scintillation counting. Protocols are shown for preparation of such standards in vitro using recombinant PI3K. Second, samples can be prepared from cells treated with a phosphoinositide 3-kinase inhibitor such as wortmannin (100 nM) or LY294002 (20 µM), which will reduce significantly all 3-OH phosphorylated lipids (16, 17).

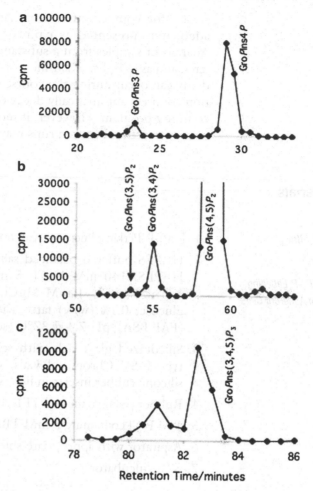

Fig. 4. Close up views of chromatographs from [^{32}P]-P$_i$ labelling showing PtdIns$P$ (**a**), PtdInsP_2 (**b**), and PtdInsP_3 (**c**). Note that PtdIns5P cannot be resolved from PtdIns4P on this system.

Scintillation data can be presented as either counts per minute (c.p.m.), or disintegrations per minute (d.p.m.). C.p.m. is the average number of counts acquired per minute by the scintillation counter, whereas d.p.m. is the average number of radioactive disintegrations in each sample per minute and is calculated from c.p.m. If counting efficiency is 100% then c.p.m. and d.p.m. will be equal, which is almost the case for [^{32}P]; however, counting efficiency is typically 30% or less for [^{3}H]. Additionally in certain scintillation fluids, the high phosphate content of samples from the later part of the gradient can reduce counting efficiency of [^{3}H] further, an effect called *quenching*. Modern scintillation fluids designed for high phosphate samples experience less quenching than older formulae; however, if accurate d.p.m. measurements need to be determined a quench correction program should be set up for your counter. For many applications, data can be adequately presented as c.p.m. So this is often unnecessary. Analysis of samples from a substantial experiment can take a few days, and as [^{32}P] has a half-life of 14.2 days significant radioactive decay can occur during the course of the analysis. Most scintillation counters automatically decay correct to the start time of the counting program. However, if more than one program is used, decay correction between runs may be necessary.

2. Materials

2.1. Labelling Protocols

2.1.1. [^{32}P]-P$_i$ Labelling of PPIn of Adherent Cells

1. [^{32}P]-P$_i$ Perkin Elmer, e.g., cat. no. NEX054010MC (see Note 1).
2. HEPES buffered balanced salts solution (HBBSS): 15 mM HEPES, 140 mM NaCl, 5 mM KCl, 2.8 mM NaHCO$_3$, 1.5 mM CaCl$_2$, 1 mM MgCl$_2$, 0.06 mM MgSO$_4$, 5.6 mM glucose, 0.1% (w/v) fatty acid-free bovine serum albumin (FAF BSA); pH 7.4 @ 37°C (see Note 2).
3. Siliconized glass vials with solvent resistant stoppers, e.g., type 5-SV Chromacol vials with type 12-SC-ST2 PTFE/silicone rubber inserts in lids (see Note 3).
4. Rubber policeman, or PTFE scraper.
5. 1 M HCl containing 5 mM TBAS, and MeOH, both ice cold.
6. Aspirator with appropriate shielding.
7. 37°C incubator.

2.1.2. [^{32}P]-P$_i$ Labelling of PPIn of Suspension Cells

1. [^{32}P]-P$_i$ Perkin Elmer, e.g., cat. no. NEX054010MC (see Note 1).
2. HBBSS: 15 mM HEPES (Sigma), 140 mM NaCl, 5 mM KCl, 2.8 mM NaHCO$_3$, 1.5 mM CaCl$_2$, 1 mM MgCl$_2$, 0.06 mM MgSO$_4$, 5.6 mM glucose, 0.1% (w/v) FAF BSA; pH 7.4 @ 37°C (see Note 2).
3. Siliconized glass vials with solvent resistant stoppers, e.g., type 5-SV Chromacol vials with type 12-SC-ST2 PTFE/silicone rubber inserts in lids (see Note 3).
4. Solvents: dH$_2$0, MeOH, CHCl$_3$.
5. Orbital shaker at 37°C.

2.1.3. [^{3}H]-Inositol Labelling of PPIn of Adherent Cells

1. Inositol free culture medium (DMEM etc.), which is available from most media suppliers, buffered to appropriate pH with 20 mM HEPES (see Note 4).
2. FAF BSA.
3. Dialysed FBS (see Note 5).
4. [^{3}H]-inositol, (e.g., GE Healthcare cat. no. TRK317).

5. Siliconized glass vials with solvent resistant stoppers, e.g., type 5-SV Chromacol vials with type 12-SC-ST2 PTFE/silicone rubber inserts in lids (see Note 3).

6. Rubber policeman, or PTFE scraper.

7. 1 M HCl containing 5 mM TBAS, ice-cold.

8. MeOH, ice-cold 9. Aspirator.

2.1.4. [³H]-Inositol Labelling of PPln in Suspension Cells

1. Inositol free medium (e.g., RPMI available from Gibco).

2. FAF BSA.

3. Dialysed FBS (see Note 5).

4. [³H]-inositol, e.g., GE Health Care cat. no. TRK317.

5. Siliconized glass vials with solvent resistant stoppers, e.g., type 5-SV Chromacol vials with type 12-SC-ST2 PTFE/silicone rubber inserts in lids (see Note 3).

2.2. Extraction of Lipids from Adherent and Suspension

1. Batch of "clean" upper and lower phases from a Folch solvent mixture prepared freshly for each extraction by mixing $CHCl_3$, MeOH, and aqueous solution (25 mM EDTA, 5 mM TBAS, 1 M HCl) in the ratio 8:4:3 (see Note 6). After mixing the phases should be separated by centrifugation at $3,000 \times g$, the upper and lower phases removed and stored in a stoppered glass bottle until use.

2. Siliconized glass vials with solvent resistant stoppers, e.g., type 5-SV Chromacol vials with type 12-SC-ST2 PTFE/silicone rubber inserts in lids (see Note 3).

3. 2.4 M HCl, 5 mM TBAS. $CHCl_3$.

4. Rotary evaporator (see Note 7).

5. Mixed brain phosphoinosides (Sigma P6023).

2.3. Lipid Deacylation and Preparation for HPLC

2.3.1. Preparation of Monomethylamine Reagent

1. Monomethylamine gas (Sigma/Fluka cat. no. 65571), 175 g cylinder with tap fitting.

2. Methanol.

3. n-butanol.

4. dH_2O.

5. Dry ice.

6. Fume hood.

7. Glass vials for storage, e.g., 20 mL glass scintillation vials.

2.3.2. Deacylation of Lipid Samples and Preparation for HPLC Analysis

1. Monomethylamine reagent, prepared as below (Fluka) (see Note 8).

2. Water bath at 53°C.

3. Rotary evaporator (see Note 7).

4. Butanol.

5. Petroleum-ether (40–60 fraction).

6. Ethyl formate, dH_2O.

7. 0.1 mM EDTA.

8. 0.45 μM luer lock filters (Millipore, type HV; cat. no. SJHV004NS).

2.4. HPLC Analysis of Samples

1. HPLC system capable of generating programmable gradients comprising two mobile phases with a manual injector (Rheodyne 7725) and 5 mL sample loop (see Note 9). In line degassing unit is optional.

2. Whatman Partisphere SAX, 4.6×125 mm column (see Note 10).

3. 2.5 M NaH_2PO_4, 0.22 μm filtered and degassed (see Note 11).

4. dH_2O, 0.22 μm filtered and degassed (see Note 11).

5. 0.45 μM luer lock filters (Millipore, type HV; cat. no. SJHV004NS).

6. 2 mL luer lock syringes.

7. Luer lock injection needle.

8. Programmable fraction collector capable of collecting between 170 and 180 samples in 5 mL vials.

9. Scintillation counter.

10. 5 mL scintillation vials.

11. Scintillation fluid capable of solubilizing samples containing high levels of phosphate, e.g., Packard UltimaFlow AP (see Note 12).

2.5. Preparation of Labelled Standards

2.5.1. Preparation of [³²P]-Labelled Standards

1. Recombinant PI3K, e.g., PI3K alpha Jena Biosciences cat. no. PR-335 (see Note 13).

2. Substrate PtdIns (e.g., Sigma cat. no. 79401).

3. PtdIns4P (e.g., Calbiochem cat. no. 524647).

4. 4.PtdIns(4,5)P_2 (e.g., Calbiochem cat. no. 524644).

5. PtdIns5P (CellSignals Inc).

6. PtdEtn (e.g., Sigma P7943), or mixed brain PPIn (Sigma P6023), (see Note 14).

7. 0.1 M $MgCl_2$.

   (a) mM ATP.

8. Kinase buffer: 20 mM HEPES pH 7.4, 120 mM NaCl, 0.1 mM EDTA, and 1 mM DTT.

9. (^{32}P)-γATP, e.g., Perkin Elmer cat. no. NEG502A (see Note 1).

10. Screw capped 1.5 mL microfuge tubes (see Note 15).

11. Methanol.

12. Chloroform.

13. 2.4 M HCl with 5 mM TBAS.

14. 1 M HCl with 5 mM TBAS.

15. Mixed brain phosphoinosides (Sigma P6023).

16. Rotary evaporator (see Note 7).

17. Bath sonicator, e.g., Grant XB2.

2.5.2. Preparation of [³H]-Labelled Standards

1. Recombinant PI3K, e.g., PI3K alpha Jena Biosciences cat. no. PR-335 (see Note 13).

2. [³H]-labelled substrate: [³H]-PtdIns Perkin Elmer cat. no. NET862; or [³H]-PtdIns(4,5)P_2 Perkin Elmer cat. no. NET895.

3. PtdEth (e.g., Sigma P7943), (see Note 14).

4. Unlabelled substrate: PtdIns4P (e.g., Calbiochem Cat. No. 524647), PtdIns(4,5)P_2 (e.g., Calbiochem cat. no. 524644), PtdEtn (e.g., Sigma P7943), (see Note 14).

5. Solutions: 0.1 M $MgCl_2$; 100 mM ATP; kinase buffer 20 mM HEPES pH 7.4, 120 mM NaCl, 0.1 mM EDTA, and 1 mM DTT.

6. Screw capped 1.5 mL microfuge tubes (see Note 15).

7. Methanol.

8. Chloroform.

9. 2.4 M HCl with 5 mM TBAS.

10. 1 M HCl with 5 mM TBAS.

11. Mixed brain phosphoinosides (Sigma P6023).

12. Rotary evaporator (see Note 7).

13. Bath sonicator, e.g., Grant XB2.

3. Methods

3.1. Labelling Protocols

3.1.1. [³²P]-P$_i$ Labelling of Adherent Cells

1. Cells should be passaged into an appropriate petri-dish so that they are sub-confluent at the time for the experiment. 6 cm dishes provide a satisfactory level of labelling for most applications. If the cells are to be deprived of serum, this should be done prior to labelling (see Note 16).

2. The cells should be washed twice into HBBSS prewarmed to 37°C and incubated in 0.5 mL HBBSS containing 0.1% FAF BSA (w/v), 0.3 mCi/mL [³²P]-P$_i$ for 1 h at 37°C.

3. After 1 h of labelling, aspirate off the labelling medium and wash the cells twice with HBBSS (see Note 17).

4. The cells can now be manipulated according to the experimental protocol employed. For example, cells can be stimulated with

an agonist by adding medium exchanging the HBBSS on the cells with buffer containing various agonists.

5. The cells are killed for the removal of HBBSS with an aspirator and addition of 0.5 mL ice cold 1 M HCl, 5 mM TBAS and placed on ice (see Note 18).

6. Cells should be harvested by scrapping with a PTFE scrapper/ rubber policeman and decanting into a clean glass vial.

7. The petri dish and scrapper should be washed with 0.667 mL of ice cold MeOH, and the MeOH combined with the initial 1 M HCl wash (see Note 19).

8. As the samples contain 1 M HCl, it is not appropriate to store the cells and lipid extraction should be undertaken as soon as possible after killing the cells.

3.1.2. [^{32}P]-P$_i$ Labelling of Suspension Cells

1. The cells to be labelled were washed twice and resuspended in HBBSS at a density of $2.5.10^7$/mL with 0.3 mCi/mL [^{32}P]-P$_i$ in a siliconized glass vial.

3. The cells should be incubated in an orbital shaker, at 37°C for 1 h.

4. The labelled cells should be washed three times in HBBSS and resuspended in HBBSS at a density of $2.5.10^7$/mL.

5. 150 µl aliquots of [^{32}P]-P$_i$-labelled cells (see Note 20) can be challenged with agonist in siliconized glass vials. Usually this is done by the addition of agonist in 20 µl of HBBSS. All subsequent procedures for nonadherent cells will assume a final volume of 170 µl of cells.

6. Cells should be killed by the addition of 750 µl of a solution of CHCl$_3$, MeOH, and water to give a final ratio of 1:3.75 (v/v) H$_2$O:MeOH/CHCl$_3$ (2:1, v/v), which will give a one phase system. Typically this meant an addition of 750 µl of a solution of 96.8% (v/v) MeOH/CHCl$_3$ (2:1, v/v), 3.14% (v/v) H$_2$O.

7. The extracts can be stored at −70°C or processed immediately.

3.1.3. [^{3}H]-Ins Labelling of Adherent Cells

1. Cells should be passaged into an appropriate petri-dish so that they are sub-confluent at the time for the experiment. 6 cm dishes provide a satisfactory level of labelling for most applications (see Note 16).

2. The cells should be washed twice into inositol-free medium containing 0.1% dialysed FBS and 1–20 µCi/mL [^{3}H]-inositol.

3. Cells should be incubated for upto 48 h (see Note 23).

4. The cells can now be manipulated according to the experimental protocol employed. For example, cells can be stimulated with an agonist by adding medium exchanging the HBBSS on the cells with buffer containing various agonists.

5. The cells are killed for the removal of HBBSS with an aspirator and addition of 0.5 mL ice cold 1 M HCl with 5 mM TBAS solution and placed on ice (see Note 18).

6. Cells should be harvested by scrapping with a PTFE scrapper/rubber policeman and decanting into a clean glass vial.

7. The petri dish and scrapper should be washed with 0.667 mL of ice cold MeOH, and the MeOH combined with the initial 1 M HCl wash (see Note 19).

8. As the samples contain 1 M HCl, it is not appropriate to store the cells and lipid extraction should be undertaken as soon as possible after killing the cells.

3.1.4. [³H]-Ins Labelling of Suspension Cells

1. Cells were washed once with inositol-free medium (see Note 21) supplemented with 0.1% (w/v) FAF BSA, and resuspended at 1.10^5/mL in inositol-free medium, supplemented with 5% (v/v) dialysed FBS and containing 1–20 µCi/mL [³H]-inositol (see Note 22).

2. Cells were incubated with [³H]-inositol for up to 48 h prior to the experiment (see Note 23).

3. Cells were washed twice in HBBSS containing 0.1% (w/v) FAF BSA and resuspended at a density of $2.5.10^7$/mL.

4. Cells can now be challenged/stimulated and killed as described for [³²P]-P$_i$-labelling from step 5.

3.2. Extraction of Lipids

3.2.1. Extraction of Lipids from Suspension Cells

1. To cells labelled and killed as described above add 170 µl of 2.4 M HCl with 5 mM TBAS solution and 725 µl of CHCl$_3$ containing 20 µg of mix brain PPIn.

2. Mix the samples by vortexing and centrifuged at $3,000 \times g$ for 5 min. The samples will separate into two phases, with approximately 1,100 µl of lower phase and 710 µl of upper phase. Process all the samples in parallel.

3. Remove the lower phase and place in a second tube containing 710 µl of clean upper phase from "clean" Folch phase split (see Note 24), being careful not to carry over any protein from the interface between the two phases.

4. Add 1,100 µl of clean lower phase added to original tube.

5. Both the original set of tubes and the second set of tubes should be mixed thoroughly and centrifuged at $3,000 \times g$ for 5 min.

6. Remove the lower phase from the second tube and place in a clean third tube. Then using the same pipette tip remove the lower phase from the first tube and place in the second tube. Discard the first tube.

7. Mix by vortexing both the second set of tubes and centrifuge at $3,000 \times g$ for 5 min.

8. Transfer the lower phase from the second tube to the third tube. Discard the second tube.

9. Dry down all samples in a rotary evaporator (see Note 7).

10. If necessary samples can be stored at –70°C overnight before deacylation and subsequent analysis.

3.2.2. Extraction of Phospholipids from Adherent Cells

1. To the samples prepared as described above add 1.33 mL of $CHCl_3$ containing 20 μg of mix brain PPIn, mix by vortexing and centrifuge at $1,000 \times g$ for 5 min. The phases should separate to give approximately 1 mL of upper phase and 1.5 mL of lower phase.

2. The extraction procedure is identical to that given for nonadherent cells, except that the volumes of lower and upper phases are 1 mL and 1.5 mL, respectively.

3.3. Lipid Deacylation and Preparation for HPLC

3.3.1. Preparation of Monomethylamine Reagent

1. In a fume hood, monomethylamine gas (Fluka), from a 175 g cylinder, should be bubbled slowly through 268 mL of a mixture of 4:3:1 (v/v/v) methanol/H_2O/n-butanol, on dry ice, until the volume of the mixture has expanded to 465 mL (see Note 25).

2. The methylamine reagent was aliquoted into clean glass scintillation vials, which were closed securely, and stored at –70°C until used (see Note 25).

3.3.2. Deacylation of Lipids and Preparation for HPLC Analysis

1. Add 250 μl of monomethylamine reagent to the lipid samples prepared as described above, and mix the samples by vortexing.

2. Incubate the samples at 53°C for 30 min.

3. Let samples cool to room temperature and centrifuge at ≈$3,000 \times g$ for 5 min.

4. Dry the samples in vacuo in a rotary evaporator (see Note 7).

5. To the dried samples at 0.5 mL of dH_2O, and 0.6 mL of butanol/petroleum-ether (40–60 fraction)/ethyl formate (20:4:1, v/v/v).

6. Mix the samples by vortexing and centrifuge at ≈$3,000 \times g$.

7. Remove the upper organic phase and discard.

8. Add another 0.6 mL of butanol/petroleum-ether (40–60 fraction)/ethyl formate (20:4:1, v/v/v) and repeat steps 6 and 7.

9. Dry the samples in vacuo.

10. Resuspend the samples in 0.5 mL of sterile 0.1 mM EDTA.

11. Remove a 10 μl aliquot from each sample in a scintillation vial, add scintillation fluid and count radioactivity. This will give a good indication as to whether the labelling and extraction procedure has been reproducible.

12. Store samples at –70°C until analysis.

3.4. HPLC Analysis of Lipids Head Groups

3.4.1. Running Samples on HPLC

1. The HPLC pumps should be primed with the appropriate solvents.

2. If a new column is attached to the HPLC system, then the flow rate should be set to 0.1 mL/min and filtered dH_2O is pumped through the column for at least 30 min. New columns are stored in methanol and pumping salt into the column might cause precipitation and fouling of the column.

3. With the column in line, the flow rate should be increased from 0 to 1 mL/min over 10 min.

4. The HPLC system used should be set up to run the gradient shown in Table 2. For best reproducibility a gradient containing no sample should be run prior to loading the first sample in a batch of runs. Each run will be for 120 minutes duration.

Table 2
HPLC gradient used to separate Gro-Ins*P* head groups. Buffer A is dH_2O and buffer B is 2.5 M NaH_2PO_4. The flow rate is 1 mL/min. Gradient includes a washing step, and a new sample should be loaded immediately after the end of the run. This gradient was developed for use with a 4.6 × 125 mm Whatmann Partisphere SAX column (11), and the author has used this gradient successfully without modification for more than 10 years. If other columns are used the gradient might need adjusting, for example, if a 250 mm column is used then a longer washing step will be required

Time (min)	%A	%B
0	100	0
1	99	1
30	93	7
31	85	15
60	71	29
61	67	33
80	40	60
81	0	100
85	0	100
86	100	0
120	100	0

5. The samples to be analysed should be thawed and filtered through a 0.45 µm PVDF filter. Any internal standards should be added at this point (see Note 26). The original tube containing the sample should be washed with 1 mL dH$_2$0, and this is filtered into the sample through the same filter.

6. Samples should be loaded onto a 5 mL sample loop, injected onto the column, and the gradient and fraction collector started. This can often be set up to be triggered from the injector. Fractions should be collected every 30 s for the full duration of the gradient (90 min).

7. After the run has finished, 2 mL scintillation fluid should be added to the samples and the samples counted in a scintillation counter. The time of counting per sample will be determined by the number of counts in each fraction (see Note 27).

8. After 120 min the next sample can be chromatographed.

9. After a batch of samples have been analysed the HPLC should be turned off by slowly reducing the flow rate from 1 mL/min to 0 over 10 min. This gradual reduction in flow rate avoids cavitation in the column.

10. Typical chromatographs for [^{33}P]-P$_i$ and [^{3}H]-inositol labelling are shown in Figs. 3 and 4, and typical retention times are shown in Table 3.

Table 3
Approximate retention times of various PPIn. Times will vary slightly between different batches of column and with ageing of the column. Generally the retention times of PtdInsP_2 isomers are the most consistent. After fitting a new column the retention times will vary for the first 10 or so runs, after which the column should operate consistently. Towards the end of the column's useful life retention times can vary and resolution will be lost

PPIn	Retention time (minutes)
PtdIns	8–12
PtdIns3P	22–24
PtdIns4P	28–30
PtdIns(3,5)P_2	53
PtdIns(3,4)P_2	55
PtdIns(4,5)P_2	60
PtdIns(3,4,5)P_3	75–80

3.4.2. Analysis of Data

1. A background count is estimated from the counts in blank vials and this subtracted from each vial. Many counters can be set up to do this automatically.

2. Once peaks containing the various GroPIns have been identified the total counts in those peaks are added together.

3. Data for [^{33}P]-P$_i$ labelling may need decay correcting to a specific time point as the decay (approximately 5% per day) will be significant while analysing a number of samples.

4. An example of the type of data that can be produced is shown in Table 4.

3.5. Preparation of Labelled Standards

3.5.1. Preparation of [^{32}P]-Labelled Standards

1. Aliquot substrates and dry in vacuo in a microfuge tube. Use 100 µM final concentration of substrate, which for a 65 µl reaction volume equates to 5.6 µl/per reaction of a 1 mg/mL solution. PtdIns4P, PtdIns5P, and PtdIns(4,5)P$_2$ should be presented with 400 µM PtdEtn, which corresponds to 2.5 µl/reaction of a 10 mg/mL solution. PtdIns and mixed brain phosphoinositides do not require addition of PtdEtn.

Table 4
Typical data from experiment looking at PtdIns(3,4,5)P$_3$ accumulation in U937 cells. U937 cells were serum starved for 16 h, [^{32}P]-P$_i$ labelled and challenged with various agonists or vehicle (HBBSS) as described. Labelled lipids were extracted and PtdIns(3,4,5)P$_3$ measured by HPLC. Error bars represent the range of the data (*n* = 2). Agonists used were: ATP (100 µM, 40 s); Insulin (10 µg/mL, 40 s); interleukin-8 (1 µg/mL, 40 s); Leukotriene D$_4$ (LD4, 500 nm, 15 s); Bradykinin (Bk, 10 µM, 40 s); Interleukin-6 (IL-6, 1,000 U/mL, 40 s); IGF-I (10^{-8} M, 40 s)

	PtdInsP_3	Error
Control	5,270	±707
ATP	16,846	±280
Insulin	18,202	±1,268
IL-8	8,811	±383
LD4	7,916	±182
Bk	5,646	±40
IL-6	9,645	±476
IGF-1	20,299	±1,067

2. Sonicate lipids into 42.5 µl of kinase buffer per reaction using a bath sonicator (see Note 28), and add 1 µl (1 µCi) of [^{32}P]-γATP and 2.5 µl of a 1 in 46 dilution of 100 mM ATP (to give 50 µM final) in kinase buffer per reaction, i.e., for five reaction you would sonicate the lipids into 212.5 µl of kinase buffer and add 5 µl of [^{32}P]-γATP and 12.5 µl of a 1 in 46 dilution of 100 mM ATP in kinase buffer. Leave the lipid/ATP mixture at room temperature until use.

3. Thaw the PI3K on ice (see Note 29) and for each reaction add 1 µl (0.2 µg of Jena Biosciences PR-335) to 19 µl of an ice-cold solution of 5.12 mM $MgCl_2$ (1 in 19.5 dilution of 100 mM $MgCl_2$ in kinase buffer) in a screw capped microfuge tube.

4. Start reaction by adding 45 µl of the lipid substrate/ATP mix to the enzyme and incubate at room temperature for 20 min (see Note 30).

5. Stop reaction by the addition of 243 µl of 2:1 (v/v) MeOH/$CHCl_3$.

6. Split the phases by adding 56.5 µl 2.4 M HCl, 5 mM TBAS, and 243 µl $CHCl_3$ containing 20 µg of mixed brain phospholipids.

7. The samples should be mixed by vortexing and centrifuged at full speed in a microcentrifuge for 5 min.

8. Remove the lower phase carefully and transfer to a fresh tube containing 237 µl of clean upper phase. Clean upper phase is made by mixing the 1 M HCL with 5 mM TBAS solution together with MeOH, $CHCl_3$, in the ratio 47:48:3.

9. Mix the samples by vortexing and centrifuge at full speed in a microcentrifuge for 5 min.

10. The lower phase should be removed into a fresh microcentrifuge tube and dried in vacuo.

11. The [^{32}P]-labelled standards should be deacylated as described above and stored at –70°C in 0.1 mM EDTA until needed.

3.5.2. Preparation
of [^{3}H]-Labelled Standards

1. Aliquot substrates and dry in vacuo in a microfuge tube. Use 100 µM final concentration of unlabelled substrate, which for a 65 µl reaction volume equates to 5.6 µl/per reaction of a 1 mg/mL solution, and 1–10 µCi of labelled substrate (see Note 31). PtdIns(4,5)P_2 should be presented with 400 µM PtdEtn, which corresponds to 2.5 µl/reaction of a 10 mg/mL solution.

2. Reactions should be performed as described for [^{32}P]-standard generation, except that no [^{32}P]-γATP should be included in the assay.

4. Notes

1. GE Health Care (formally Amersham/Pharmacia) have in their infinite wisdom stopped supplying [^{32}P]-labelled compounds. Despite several attempts the author has had no satisfactory explanation from GE Health Care as to why this decision has been made; however, the suspicion must be that some bean counter did not think it was profitable enough. This leaves Perkin Elmer as the sole supplier of [^{32}P]-P$_i$ and [^{32}P]-γATP. Of course [^{33}P] can be used in these protocols with a corresponding reduction in ionising radiation hazard and increase in cost.

2. FAF BSA is used to avoid inadvertently stimulating cells with lyso-phosphatidic acid that is a contaminant of BSA.

3. The author now uses siliconized 5 mL Chromacol vials for most experiments. However, some of the caps supplied do not always fit the vials well, so it is advisable to check that the caps fit the vials *before* any radioactive material is put into the vial. To reduce lipid binding, vials should be siliconized with either Repelcote or Sigmacoat according to the manufacturer's instructions and washed thoroughly.

4. When handling small volumes of culture medium outside a 5/10% CO_2 environment, it is advisable to add buffer to avoid the pH change caused by de-gassing.

5. FBS contains significant quantities of inositol, which can be removed by dialysis (see Chapter 1). Most suppliers sell dialysed FBS.

6. The author makes a clean Folch split by adding together (in this order), 7.09 mL 1 M HCl, 375 μl 0.5 M EDTA, 37.5 μl 1 M TBAS, 10 mL MeOH, and 20 mL CHCl$_3$ in a 50 mL polypropylene Falcon tube. This mixture should be mixed and centrifuged at 3,000×g. By adding the organic solvents last evaporation is limited. Once the phases have been separated the upper and lower phases should be immediately transferred to stoppered glass bottles. It is important to add EDTA to the phase split, as a small amount of this will carry over into the lipid extract and chelate metal ions in the deacylation step, thus improving the fidelity of this reaction.

7. Care should be taken when choosing a rotary evaporator, as the protocols described use both aggressive chemicals (HCl and monomethylamine), which can corrode pump bearings and volatile solvents (CHCl$_3$, and petroleum ether), which will pass through a standard −80°C trap and dissolve into the pump's lubricating oil thus reducing its effectiveness. Both these situations will eventually destroy a standard vane

pump (e.g., Edwards style). The two options are to have an expensive −120°C condensing trap, or to employ a PTFE diaphragm pump and a −80°C trap. As PTFE diaphragm pumps are resistant to aggressive chemicals, any which pass through the trap will not corrode the pump mechanism. The pump outlet should of course be vented into a fume cabinet. The author uses a UNIVAPO 100 H rotary evaporator that can accommodate 5 mL Chromacol vials in a 40-13 rotor, or microcentrifuge tubes in a 72-11 rotor, and a Laboport cat. no. N810.3FTP PTFE diaphragm pump.

8. The author has always used methylamine reagent prepared as described (18). There are commercial preparations of methylamine reagent available; however, several of these use water as a solvent – not ideal for deacylating lipids. If a commercial preparation of methylamine reagent is used it might be worth adding methanol and butanol to give the same proportions as that given above.

9. As samples are concentrated on a SAX column, there is no need to load the samples in as small a volume as possible, so for better reproducibility a 1.5 mL sample volume is suggested. However, due to the very small bore of HPLC tubing, there is a significant boundary layer that reduces the effective volume of the sample loop. Thus, it is best to err on the side of caution and have a samples loop at least twice the volume of the sample.

10. The author has used 4.6×125 mm Whatman Partisphere SAX columns (cat no. 4621-0505) for over 14 years now and has generally had good results with them. A "good" column will give acceptable performance for over 100 HPLC runs. Some batches of columns seem to work better (and for longer) than others so it is worth recording batch numbers for future reference. The widespread use of Whatman Partisphere SAX columns is largely historical as this was the first column of its type used for PPIn separation. It is likely that there are now better columns than the Whatman on the market; however, if other columns are used then it is probable that the gradient shown in Table 2 will have to be modified.

11. Mobile phases should be degassed prior to use in HPLC. The main reason for doing this is to avoid air bubbles being trapped in the pump heads, which can cause irreproducible gradients. Solvents can be degassed by sparging with helium, which is wasteful; under vacuum; by boiling, or by use of an online degassing unit. The author combines degassing under vacuum with filtering by using disposable 22 μm bottle filters, and uses a degassing unit inline between the solvent reservoirs and the pump heads.

12. It is important to use a scintillation fluid that is designed to solubilize samples containing high levels of phosphate. Most scintillation fluids are unable to do this. Packard Ultima Flow AP is designed specifically for this purpose. Sadly it is very expensive when compared with other fluids.

13. Recombinant PI3K can be purchased from several commercial sources. The author has used PI3K cat. no. PR-335 from Jena biosciences with good results.

14. In vitro PI3K will phosphorylate all PPIn without a D-3 phosphate. Obviously, the lipid substrates used should be appropriate for the standard needed, e.g., PtdIns$(4,5)P_2$ should be used as substrate to produce a $[^{32}P]$-PtdIns$(3,4,5)P_3$ and mixed brain PPIn to make a mixture of $[^{32}P]$-labelled D-3 GroPInsP_n. PtdIns, PtdEth, and mixed brain phosphoinositides should be dissolved in CHCl$_3$ to 5, 10, and 10 mg/mL respectively. PtdIns4P and PtdIns5P should be dissolved in 2:1 (v/v) CHCl$_3$/MeOH to 1 mg/mL. PtdIns$(4,5)P_2$ can be dissolved in 2:1 (v/v) CHCl$_3$/MeOH to 1 mg/mL; however, a very small quantity of water may be needed to be added dropwise to get PtdIns$(4,5)P_2$ into solution. PtdIns$(4,5)P_2$ stores poorly in solution due to the susceptibility of its phosphates to hydrolysis, so for long-term storage it may be wise to aliquot and lyophilize. All lipid solutions should be stored at –70°C in sealed glass bottles. The author uses Chromacol vials for this and uses Parafilm on the caps as extra insurance.

15. The author recommends use of screw-capped microfuge tubes with an o-ring in the cap. Although these are less convenient to use, the improved sealing prevents leakage of radioactivity.

16. Mammalian cells vary considerably in their tolerance to serum starvation, and this should be investigated thoroughly prior to embarking on a labelling experiment. It should also be noted that ensuring an equal number of cells in each dish is important for reproducibility, and that this is often an area where improvements could be made.

17. Care should be taken to minimize the loss of cells from the petri dish during washing.

18. A convenient way of keeping the cells ice cold is to have an aluminium tray in an ice bucket and place the cells on the tray. The tray will provide a flat surface for stability, as the samples will be quite radioactive.

19. It goes without saying that this step presents a great opportunity to cause significant radioactive contamination of the surrounding area. Furthermore, the reproducibility of the experiment replies on a high recovery of material from this

stage. Both the criteria mean that great care should be exercised in this step of the procedure.

20. 150 µl aliquots of [^{32}P]-P$_i$-labelled cells labelled in this fashion are usually sufficient for adequate detection of D-3 PPIn.

21. Mammalian cells vary in their tolerance to inositol-free medium, especially when combined with serum starvation. A trial of different periods of inositol/serum withdrawal should be done prior to any labelling experiments (see Chapter 1).

22. The amount of label used in each experiment will depend on the ability of the cells to take up label and the sensitivity required.

23. Although labelling to equilibrium might be the aim, the time of incubation with label might have to be adjusted if the cells being used are not able to tolerate withdrawal of inositol/serum from the medium. For example, U937 cells only really maintain viability for about 16 h of serum/inositol withdrawal. Alternatively, a small amount of inositol can be added to the medium (12).

24. When pipetting volatile organic solvents it is important to equilibrate the pipette with the solvent before use. This prevents evaporating solvent in the pipette ejecting solvent from the tip. To equilibrate the tip, pipette some CHCl$_3$ up and down into each new tip.

25. Monomethylamine reagent prepared in this fashion is effective for several years when stored at –70°C.

26. Internal standards generated as outlined in this chapter, and labelled with a different isotope to the sample can be added. One problem with this approach is when the samples are counted [^{32}P] signal will bleed into the ^{3}H window. This problem can be avoided when using [^{3}H]-labelled sample and [^{32}P]-labelled standards by loading a sufficiently low amount of [^{32}P]-labelled standard. It is a good idea to HPLC analyse an aliquot of the [^{32}P]-labelled standard before adding this to the [^{3}H]-sample. Alternatively, a program can be set up on the scintillation counter to compensate for any bleed through between the two channels.

27. Counting error is proportional to the square root of the total counts seen. Thus, improved accuracy can be achieved by increasing the counting time. Most scintillation counters will automatically calculate counting error. The end user must thus decide what is deemed acceptable.

28. Sonication efficiency is improved by adding some detergent such as Decon 75 to the water in the sonication bath. In the author's experience lipids should sonicate easily to give a slightly turbid solution. If this does not occur, then there is likely to be a problem with the substrate.

29. Lipid kinases are relatively robust enzymes and retain their activity for many months when stored at −70°C; however, it is good practice to avoid repeated freeze thaw cycles and so the PI3K should be frozen in conveniently sized aliquots after its initial thawing.

30. For accurate kinetic measurements assays should be incubated at a fixed temperature in a water bath or heating block. 30°C is usually chosen for these type of experiments. However, for the purposes of making lipid standards room temperature incubation is fine, and this allows the use of a Perspex block to reduce exposure to ionizing radiation. Longer incubation times will give higher incorporation, although the author finds 20 min more than adequate.

31. Labelled substrate is expensive, so the total turnover should be maximized. Longer incubation time would be one way of achieving this.

References

1. Parker, P. J. (2004) The ubiquitous phosphoinositides. *Biochem. Soc. Trans.* **32**, 893–898.

2. Michell, R. H. (1975) Inositol phospholipids and cell surface receptor function. *Biochim. Biophys. Acta* **415**, 81–147.

3. Whitman, M., Downes, C. P., Keeler, M., Keller, T., and Cantley, L. (1988) Type I phosphatidylinositol kinase makes a novel inositol phospholipid, phosphatidylinositol-3-phosphate. *Nature* **332**, 644–646.

4. Gray, A., Olsson, H., Batty, I. H., Priganica, L., and Downes, C. P. (2003) Nonradioactive methods for the assay of phosphoinositide 3-kinases and phosphoinositide phosphatases and selective detection of signaling lipids in cell and tissue extracts. *Anal. Biochem.* **313**, 234–245.

5. Guillou, H., Stephens, L. R., and Hawkins, P. T. (2007) Quantitative measurement of phosphatidylinositol 3,4,5-trisphosphate. *Methods Enzymol.* **434**, 117–130.

6. van der Kaay, J., Batty, I. H., Cross, D. A., Watt, P. W., and Downes, C. P. (1997) A novel, rapid, and highly sensitive mass assay for phosphatidylinositol 3,4,5-trisphosphate (PtdIns(3,4,5)P_3) and its application to measure insulin-stimulated PtdIns(3,4,5)P_3 production in rat skeletal muscle in vivo. *J. Biol. Chem.* **272**, 5477–5781.

7. Milne, S. B., Ivanova, P. T., DeCamp, D., Hsueh, R. C., and Brown, H. A. (2005) A targeted mass spectrometric analysis of phosphatidylinositol phosphate species. *J. Lipid Res.* **46**, 1796–1802.

8. Pettitt, T. R., Dove, S. K., Lubben, A., Calaminus, S. D., and Wakelam, M. J. (2006) Analysis of intact phosphoinositides in biological samples. *J. Lipid Res.* **47**, 1588–1596.

9. Auger, K. R., Carpenter, C. L., Cantley, L. C., and Varticovski, L. (1989) PDGF-dependent tyrosine phosphorylation stimulates production of novel polyphosphoinositides in intact cells. *Cell* **57**, 167–175.

10. Hawkins, P. T., Jackson, T. R., and Stephens, L. R. (1992) Platelet-derived growth factor stimulates synthesis of PtdIns(3,4,5)P3 by activating a PtdIns(4,5)P2 3-OH kinase. *Nature* **358**, 157–159.

11. Stephens, L. R., Hughes, K. T., and Irvine, R. F. (1991) Pathway of phosphatidylinositol(3,4,5)-trisphosphate synthesis in activated neutrophils. *Nature* **351**, 33–39.

12. Yu, J., Berggren, P. O., and Barker, C. J. (2007) An autocrine insulin feedback loop maintains pancreatic beta-cell 3-phosphorylated inositol lipids. *Mol. Endocrinol.* **21**, 2775–2784.

13. Whiteford, C. C., Brearley, C. A., and Ulug, E. T. (1997) Phosphatidylinositol 3,5-bisphosphate defines a novel PI 3-kinase pathway in resting mouse fibroblasts. *Biochem. J.* **323**, 597–601.

14. Folch, J., Lees, M., and Sloane Stanley, G. H. (1957) A simple method for the isolation and

purification of total lipides from animal tissues. *J. Biol. Chem.* **226**, 497–509.

15. Bligh, E. G., and Dyer, W. J. (1959) A rapid method of total lipid extraction and purification. *Can. J. Biochem. Physiol.* **37**, 911–917.

16. Arcaro, A., and Wymann, M. P. (1993) Wortmannin is a potent phosphatidylinositol 3-kinase inhibitor: the role of phosphatidylinositol 3,4,5-trisphosphate in neutrophil responses. *Biochem. J.* **296**, 297–301.

17. Vlahos, C. J., Matter, W. F., Hui, K. Y., and Brown, R. F. (1994) A specific inhibitor of phosphatidylinositol 3-kinase, 2-(4-morpholinyl)-8-phenyl-4H-1-benzopyran-4-one (LY294002). *J. Biol. Chem.* **269**, 5241–5248.

18. Clarke, N. G., and Dawson, R. M. (1981) Alkaline O leads to N-transacylation. A new method for the quantitative deacylation of phospholipids. *Biochem J.* **195**, 301–306.

Chapter 13

Phosphoinositide Analysis by Liquid Chromatography–Mass Spectrometry

Trevor R. Pettitt

Abstract

The phosphoinositides are a highly dynamic group of molecules implicated in many cellular control processes; however, the analysis of many of these structures has proven very difficult and time-consuming, with limited sensitivity and/or discrimination. Recent developments in LCMS now provide the prospect of routine structural and quantitative analysis of all the known phosphoinositides (and possibly some as yet unidentified structures) at high sensitivity in any biological sample. The procedures described here give very high extraction recovery from a variety of biological matrices and enable chromatographic resolution of most phosphoinositides as their native structures. When coupled with the accurate mass and fragmentation capabilities of an MS, full structural and isomeric identification can be achieved.

Key words: Lipids, Phospholipids, Phosphoinositides, Phosphatidylinositol, Lipidomics, HPLC, LCMS, Mass spectrometry, Normal-phase

Lipid Abbreviations

PtdIns	phosphatidylinositol
PtdIns$4P$	phosphatidylinositol 4-phosphate
PtdIns$3P$	phosphatidylinositol 3-phosphate
PtdIns$5P$	phosphatidylinositol 5-phosphate
PtdIns$(4,5)P_2$	phosphatidylinositol 4,5-bisphosphate
PtdIns$(3,4)P_2$	phosphatidylinositol 3,4-bisphosphate
PtdIns$(3,5)P_2$	phosphatidylinositol 3,5-bisphosphate
PtdIns$(3,4,5)P_3$	phosphatidylinositol 3,4,5-trisphosphate

1. Introduction

Traditionally, phosphoinositide {PtdIns, PtdIns$3P$, PtdIns$4P$, PtdIns$5P$, PtdIns$(3,4)P_2$, PtdIns$(3,5)P_2$, PtdIns$(4,5)P_2$, PtdIns$(3,4,5)P_3$} analysis has been a complex and time-consuming

Christopher J. Barker (ed.), *Inositol Phosphates and Lipids: Methods and Protocols*, Methods in Molecular Biology, vol. 645,
DOI 10.1007/978-1-60327-175-2_13, © Humana press, a part of Springer Science+Business Media, LLC 2010

procedure, which was normally only possible with larger samples that could be radiolabelled. This effectively excluded primary tissues and samples which were only poorly radiolabelled or where availability was limited. Since phosphoinositides are highly metabolically active and help regulate many cellular processes, there is great interest in analysis techniques that can characterize very small amounts of these lipids as their native structures without the need for labelling or derivatization. A few reports of tandem MS/MS identification of PtdIns$P$, PtdInsP_2, and PtdInsP_3 have been published (1–4); however, these methods were unable to resolve the isomers. Furthermore, these approaches suffer from the ion suppression effects inherent to direct infusion of unfractionated, complex lipid mixtures, which can lead to a significant loss of sensitivity. Coupling HPLC to MS (LCMS) significantly reduces complexity and ion suppression at any time point, while enabling isobaric and isomeric discrimination (e.g., for PtdIns$(4,5)P_2$, PtdIns$(3,4)P_2$, and PtdIns$(3,5)P_2$). This makes LCMS particularly suited to the study of signalling and other minor lipids where separation from the quantitatively major lipid components is highly desirable.

As an extension of work on the LCMS lipidomic profiling of many different lipids (5), procedures have been developed for the identification and quantification of all the naturally occurring phosphoinositides (6). This group of lipids, particularly PtdInsP_2 and PtdInsP_3, is especially prone to losses during both extraction and analysis. Quantitative extraction requires great care; however, the procedures outlined can achieve >90% recovery, even for PtdInsP_3, while the LCMS methodology can provide full phosphoinositide profiling within a single 20 min chromatographic run. However, high sensitivity quantitative analysis, particularly of PtdInsP_3 and the individual PtdIns$P$ isomers, remains tricky; so these approaches are best treated as a foundation for further development.

2. Materials

2.1. Equipment and Chemicals

1. Benchtop centrifuge (e.g., Beckman).
2. Ice-cooled sonicating bath.
3. Oven.
4. Glass syringes (25, 100, 250, and 1,000 μl from Hamilton, SGE, etc.).
5. 15 mL screw-capped polypropylene centrifuge tubes (e.g., from Sarstedt).
6. 1.8 mL screw-capped, silanized glass autosampler vials (e.g., from Alltech Associates or Chromacol).

7. 100 μl silanized glass limited volume vial inserts (e.g., from Alltech Associates or Chromacol).

8. Phosphoinositide standards (e.g., from Avanti Polar Lipids or Echelon Biosciences) (see Note 1) at 10 μg/mL in chloroform/methanol (2:1 v/v) or in chloroform/methanol/water (25:25:1 v/v/v) for very polar lipids such as PtdInsP_2 and PtdInsP_3 in silanized glass vials. Most lipids are stable at –20°C for months to years (stability decreases with increasing double bonds and/or phosphate groups; light slowly induces photo-oxidation of double bonds).

9. Solvents (acetonitrile, chloroform, dichloromethane, methanol, water) should be MS or other high purity grade.

10. Chemicals (ethylamine, piperidine, sodium EDTA, tetrabutylammonium hydrogen sulphate) should be analytical grade or better.

11. Ice-cold methanolic 0.25 M HCl.

12. Phase split buffer: 0.25 M HCl, 2 mM sodium EDTA, 5 mM tetrabutylammonium hydrogen sulphate (TBAHS; 1.7 mg/mL), 0.9% (155 mM; 9 mg/mL) NaCl.

13. Neutralisation wash buffer: 100 mM sodium EDTA pH6.0 and 1.5 mL 1 mM tetrabutylammonium hydrogen sulphate (TBAHS) in methanol.

14. Tube rinse mixture: chloroform/methanol/water (5:5:1).

15. Resuspension mixture: chloroform/methanol/water (49:49:2).

16. Liquid nitrogen.

17. Methanolic 0.25 M HCl/chloroform (1:1).

18. Silanization reagent: 5% dichlorodimethylsilane in toluene.

2.2. HPLC

1. Equipment: A modern ternary or quaternary, capillary, biocompatible HPLC system, with temperature control and nanolitre autosampler injection capabilities is an ideal front end for LCMS lipid analyses. However, many standard binary analytical HPLC systems (usually designed for 4.6 mm i.d. columns using 1 mL/min flow rates and 5–20 μl injection volumes) can be optimized to work with smaller columns and lower flow rates by removing all unnecessary dead volume and replumbing with 0.13 mm i.d. tubing (5). This helps minimize solvent delay and reduces peak spreading. Flow rates of 100–200 μl/min are usually optimal for efficient electrospray ionization (ESI) MS with standard ESI probes; thus it is generally best to optimize the HPLC to work efficiently with 1.0–2.1 mm i.d. columns. This has the advantage of reducing usage of potentially harmful and costly solvent. It also means that for a given mass of injected sample, peaks elute in a smaller solvent volume, thus increasing chromatographic peak height

and hence improving sensitivity. Replace metal surfaces that come into sample contact with biocompatible material (e.g., PEEK) if possible (see Note 2) since the metal chelating polyphosphoinositides and many other phosphorylated organic compounds (7) bind to stainless steel, leading to losses.

2. HPLC Column: Silica (3 μm, 1.0 mm i.d. × 150 mm). High grade, uniformly spherical silicas with low pore size, high surface area, and very low metal content (e.g., Inertsil, Kromasil, Luna silica) generally give good separation with minimal peak tailing. A 3 μm particle size usually gives better results than 5 μm, although backpressure may be higher, potentially limiting column length to 150 mm. If available, 200 or 250 mm column lengths can improve resolving capacity. 3 μm, 1.0 mm i.d. columns are currently optimal since they can be packed to almost as high efficiency (>70,000 plates/m) as the traditional 4.6 mm analytical columns yet use substantially lower flow rates. At capillary sizes (<1 mm i.d.) packing becomes technically more difficult, often resulting in lower efficiencies. Choose columns with the highest plate count available since this generally indicates the greatest lipid resolving capability.

   Ideally, column walls and the frits at each end should be made from a biocompatible material such as PEEK or titanium; however, stainless steel replacement is currently not generally feasible for 1 mm i.d. columns due to the very high packing pressures required although it is possible at 2.1 mm i.d. Binding of polyphosphoinositides to new, highly polished stainless steel surfaces is probably minimal although this is likely to increase with aging (see Note 3).

3. HPLC Solvents: *Warning!* Most organic solvents are hazardous, many being flammable and toxic so efficient vapour extraction is essential. Wear disposable gloves when handling solvents and all dispensing should be performed into glass containers, in a fume hood. Use glass pipettes and syringes for fine measurements. To prevent contamination do not pour excess solvent back into the original stock bottle. Depending on national regulations such as those in the UK, halogenated (chlorinated) solvents and mixtures have to be disposed of separately (at greater expense) from nonhalogenated solvent mixtures by registered disposal companies.

4. To minimize background noise/contamination on the MS, use fresh solvent batches (manufacturers print an expiry date on their bottle labels) and store the mixtures in brown bottles (with PTFE lids) to reduce photo-oxidation. For optimum purity, fresh solvent mixtures should be prepared every 1–2 days since significant background contamination appears very rapidly and is increased when modifiers such as ethylamine are present (see Note 4).

5. Volatile alkalis, such as ethylamine, chromatographically slow acidic lipids including the phosphoinositides by ensuring their deprotonation and, in effect, making them more polar. They also improve peak shape and aid ESI.

6. Solvent A; chloroform/dichloromethane/methanol/water (45:45:9.5:0.5) containing 15 mM ethylamine.

7. SolventB;acetonitrile/chloroform/methanol/water(30:30:32:8) containing 15 mM ethylamine.

8. Solvent C; 60 mM piperidine in methanol (see Note 5).

2.3. Mass Spectrometer

Provided good HPLC resolution of the phosphoinositides is achieved, MS detection of molecular ions for each class will be sufficient for many analyses. This can be achieved by ESI on a single quadrupole MS. However, this type of instrument cannot provide unambiguous identification of the exact fatty acid composition, only the total number of carbons and double bonds, e.g., C38:4 for a 18:0/20:4 or a 18:2/20:2 structure (an experienced lipid analyst will often be able to predict the most prevalent structure). To get exact fatty acid compositions requires controlled fragmentation on a triple quadrupole or ion-trap (IT) MS. Similarly, resolution of structural isomers such as PtdIns3P and PtdIns4P requires the controlled fragmentation capabilities of an IT (6). These types of MS give m/z identification accuracy to approximately 0.5 Da, which can cause difficulties identifying minor components against a noisy background of similar m/z values. The use of accurate mass instruments such as Time of-Flight (ToF; accurate to <5 ppm; <0.005 Da at m/z 1,000) greatly improves this mass discrimination and removes most misidentification problems. Hybrid accurate mass/controlled fragmentation instruments capable of rapid analysis on fast chromatographic time-scales (e.g., the Shimadzu IT-ToF) are probably the ideal; however, for PtdIns3P(PtdIns5P)/PtdIns4P isomeric resolution, which needs efficient MS3 ion transfer, a dedicated IT (e.g., the Bruker HCT$_{ultra}$) may have fewer compromises and thus work better with greater sensitivity. Attomole LCMS detection limits (signal to noise >5) can be achieved for some lipid structures on certain instruments but femtomole to picomole limits are more usual.

3. Methods

3.1. Lipid Extraction

Acidified extraction is essential for quantitative recovery of polyphosphoinositides; however, extended exposure to acid can cause deacylation and possibly phosphate loss and/or rearrangement. Solvent removal without prior neutralization causes substantial loss of the parent structures (6). For these reasons, all stages in the extraction prior to EDTA neutralization should be performed without delay (see Note 6). Lipid extractions can be performed in silanized glass; however, polypropylene tubes work almost as well, are more convenient, and don't risk loss of the silyl surface. Polyphosphoinositides bind untreated glass causing losses that may exceed 50%. Sample transfers should be performed using silanized glass Pasteur pipettes or syringes.

Lipid extracted from 10^6 cells will normally provide sufficient material for analysis, although investigation of very minor structures, such as PtdInsP_3, may require larger samples.

3.1.1. Extraction from Mammalian Cells

1. Transfer sample (10^6 cells) into 15 mL polypropylene centrifuge tube (e.g., Sarstedt).

2. Centrifuge (1,000 rpm, $220 \times g$, 3 min) to pellet cells.

3. Remove as much media as possible (assume pellet volume ~150 μl).

4. Add 2 mL ice-cold methanolic 0.25 M HCl (see Note 7).

5. Add internal standards (400 ng each of PtdIns, PtdIns5P, PtdIns4P, PtdIns(4,5)P_2, PtdIns(3,4)P_2, PtdIns(3,5)P_2, PtdIns(3,4,5)P_3) (see Note 1).

6. Vortex vigorously for 30 s.

7. Sonicate in ice-cooled sonicating bath for 5 min.

8. Add 4 mL ice-cold chloroform.

9. Vortex vigorously for 30 s.

10. Sonicate in ice-cooled sonicating bath for 10 min.

11. Split phases by addition of 1.5 mL 0.25 M HCl, 2 mM sodium EDTA, 5 mM tetrabutylammonium hydrogen sulphate (TBAHS; 1.7 mg/mL), and 0.9% (155 mM; 9 mg/mL) NaCl.

12. Vortex for 30 s.

13. Sonicate in ice-cooled sonicating bath for 5 min.

14. Stand on ice for 10 min.

15. Complete phase split by centrifugation (1,100 rpm, $250 \times g$, 5 min).

16. Transfer lower phase into a clean polypropylene tube containing 1.5 mL 100 mM sodium EDTA pH 6.0 and 1.5 mL 1 mM TBAHS in methanol (see Note 8); this neutralizes the acid extract.

17. Re-extract the remaining upper phase and interfacial material with 4 mL ice-cold synthetic lower phase (see Note 9).

18. Vortex vigorously.

19. Sonicate in ice-cooled sonicating bath for 5 min.

20. Stand on ice for 5 min.

21. Complete phase split by centrifugation (1,100 rpm, $250 \times g$, 5 min).

22. Combine lower phase with original lower phase.

23. Vortex combined, neutralized lower phase vigorously.

24. Complete phase split by centrifugation (1,100 rpm, 250×g, 5 min).

25. Discard upper phase and interfacial material (if residue remains on the tube walls then transfer lower phase into a clean tube).

26. Dry under a flow of nitrogen or on a vortex evaporator.

27. Carefully rinse down tube walls with 500 µl chloroform/methanol/water (5:5:1).

28. Dry again.

29. Resuspend in 30 µl chloroform/methanol/water (49:49:2).

30. Transfer into silanized conical base 1.8 mL autosampler vials (or silanized vial inserts) using a glass syringe.

31. Rinse tube with a further in 30 µl chloroform/methanol/water (5:5:1) and combine in silanized vial to give a total volume of ~60 µl.

32. Analyze quickly to minimize any losses on standing (possible slow acid hydrolysis).

3.1.2. Extraction from Tissue

1. Freeze tissue (<50 mg) in liquid nitrogen.

2. Pulverize in liquid nitrogen-cooled pulverizer.

3. Transfer frozen, pulverized sample into a 3 mL glass homogenizer.

4. Add 1 mL ice cold methanolic HCl (0.25 M).

5. Add internal standards (400 ng each of PtdIns, PtdIns5P, PtdIns4P, PtdIns(4,5)P_2, PtdIns(3,4)P_2, PtdIns(3,5)P_2, and PtdIns(3,4,5)P_3) (see Note 1).

6. Homogenize tissue until full disrupted (connective tissue may be problematic).

7. Transfer homogenate into a 15 mL polypropylene centrifuge tube using a silanized Pasteur pipette.

8. Rinse homogenizer with 2 mL methanolic 0.25 M HCl/chloroform (1:1) and combine with homogenate.

9. Rinse homogenizer with 3 mL chloroform and combine with homogenate.

10. Continue as for extraction from mammalian cells, from step 9.

3.1.3. Extraction from Yeast

1. Centrifuge sample (10^7 cells, 1,000 rpm, 220×g, 3 min) to pellet cells.

2. Discard media – remove as much as possible (assume pellet volume ~150 µl) to prevent an early phase split.

Alternatively, if total volume (cells + media) is less than 150 µl, then steps 1 and 2 can be omitted.

3. Add 1 mL ice cold methanolic HCl (0.25 M).

4. Transfer sample into a 2 mL glass homogenizer.

5. Add internal standards (400 ng each of 17:0/20:4-PtdIns, PtdIns5P, PtdIns4P, PtdIns(4,5)P_2, PtdIns(3,4)P_2, PtdIns(3,5)P_2, and PtdIns(3,4,5)P_3) (see Note 1).

6. Continue as for Extraction from tissue, from step 6.

3.2. Phosphoinositide Separation (Fig. 1)

Column: Silica (3 μ, 1.0 × 150 mm).

Gradient: 100% solvent A held for 1 min then to 40% solvent B over 1 min then to 60% solvent B over 18 min before recycling back to solvent A over 1 min then held for a further 14 min (total run time, including recycling, 35 min).

Flow: 80 μl/min.

Injection volume: 0.5 μl.

Column Oven: 15°C

Sample cooler: 5°C

Needle wash solvent: chloroform/methanol/water (49:49:2).

Postcolumn: Solvent C at 40 μl/min (see Note 5).

Depending on the column, a slightly modified, shallower gradient (e.g., 40–50% solvent B) may improve resolution of PtdInsP and PtdInsP_2 isomers (but see Note 10). Ensure sufficient needle washing to minimize sample carry-over (see Note 11).

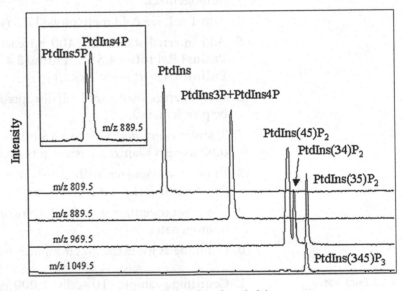

Fig. 1. LC–MS separation of 16:0/16:0 phosphoinositides. Detection as (M-H)$^-$ ions. (M-2H)$^{2-}$ and sodiated (M-2H + Na)$^-$ detected as minor ions for PtdIns$P$, PtdInsP_2, and PtdInsP_3 although under some conditions they can become predominant. Identical molecular species of PtdIns3P and PtdIns4P coelute but PtdIns5P will partially resolve (inset).

3.3. Mass Spectrometry

The settings for optimal ESI-MS sensitivity vary from instrument to instrument due to design differences in the ionizing interface. Optimum settings for phosphoinositides on a Shimadzu IT-ToF are a nebulizing (sheath) gas flow of 0.8 L/min, desolvation temperature of 270°C and probe voltage of –3.0 kV. This produces primarily singly deprotonated ions (M-H)$^-$ although doubly deprotonated (M-2H)$^{2-}$ ions and sodiated adducts (e.g., (M-2H + Na)$^-$; (see Note 12) will also be evident for the multiply phosphorylated structures. Reducing desolvation temperature increases the proportion of (M-H)$^{2-}$ (it often becomes the major ion below 200°C) but may also increase interfering background noise. This effect is highly variable and can be affected by solvent batch, previously run samples, etc.

Each additional phosphate group appears to reduce ion intensity five to tenfold (dependent on instrument) so, with the IT-ToF, the limit of detection is <1 fmol injected on-column for PtdIns, 10 fmol for PtdInsP, 100 fmol for PtdInsP$_2$ and 1 pmol for PtdInsP$_3$ although these values can be substantially poorer depending on the state of the column, the cleanliness of the MS, the purity of the running solvent, and the nature of the sample. PtdIns(3,5)P$_2$ and PtdInsP$_3$ are particularly sensitive to analysis conditions and can readily disappear. PtdInsP$_3$ is very sensitive to ion suppression by coeluting material hence a clean or well resolved sample will give a better response.

3.3.1. Isomeric Identification of PtdInsP

Full chromatographic resolution of the three known, naturally occurring PtdInsP isomers, PtdIns3P, PtdIns4P, and PtdIns5P, is probably not possible on a silica column although partial resolution of the 5-isomer from the 3- and 4-isomers (which coelute) is achievable (Fig. 2).

Quantitative determination of the individual 3- and 4-isomers is difficult but one approach that works, particularly with larger amounts of PtdInsP, uses MS3 fragmentation of the (M-2H)$^{2-}$ ions (6). The ratio of particular fragment ions is then used to determine the relative amounts of coeluting 3 (or 5) and 4 isomers from a calibration curve prepared under identical conditions using defined mixes of standard PtdInsP isomers.

Singly deprotonated (M-H)$^-$ ions do not provide sufficiently different fragmentation patterns for isomeric discrimination hence MS conditions need to be optimized for (M-2H)$^-$ formation.

1. Prepare mixtures of 100:0, 75:25, 50:50, 25:75, and 0:100 of 16:0/16:0-PtdIns3P:PtdIns4P standards (1 µg/mL total PtdInsP concentration; can use alternative molecular species).

2. Make 1 µl injections on-column using a suitable gradient and an MS method set to trap and fragment the m/z 444.2 ion to

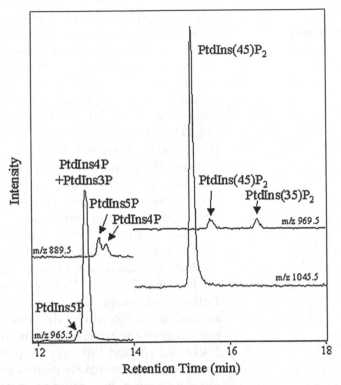

Fig. 2. Human colorectal tumor phosphoinositides. Lipids extracted with homogenization from pulverized tissue (frozen in liquid N_2). 16:0/16:0-phosphoinositides as internal standards. 16:0/16:0-PtdInsP (m/z 889.5), 18:0/20:4-PtdInsP (m/z 965.5), 16:0/16:0-PtdInsP_2 (m/z 969.5), 18:0/20:4-PtdInsP_2 (m/z 1045.5). Higher carbon number and/or more unsaturated structures elute ahead of lower carbon number, saturated structures.

MS2 then trap the resultant m/z 633.3 ion ((M-RCOO)⁻; loss of a 16:0-fatty acid chain) and fragment to MS3 (Fig. 3). The exact conditions for optimum ion transmission will have to be determined for each instrument.

3. Determine the ratio between the m/z 241 ion ((inositol phosphate–H_2O)⁻) and the m/z 409 ion (M-RCOO-224 Da)⁻ at MS3.

4. To generate the calibration curve, plot the 241/409 ratio against % PtdIns4P (Fig. 4).

This calibration appears to hold over a relatively wide range of concentrations (for best accuracy, calibrant concentrations used should be similar to PtdInsP concentrations found in the real samples) and for a range of different acyl structures although the ions used will be different. For example, with 18:0/20:4-PtdInsP, the ion at m/z 482.2 will be trapped and fragmented to give two important ions at MS2, m/z 661.3, (M-20:4)⁻ and m/z 681.3, (M-18:0)⁻, both of which are subsequently trapped and

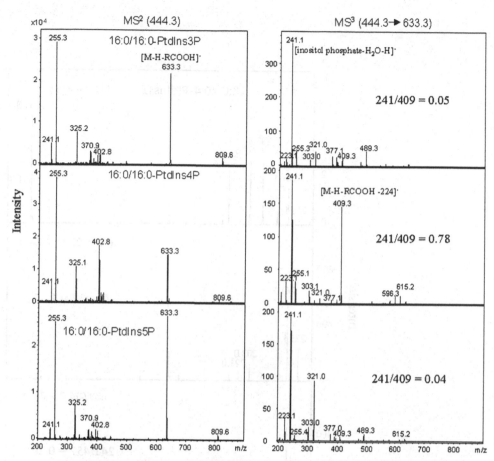

Fig. 3. Diagnostic fragmentation of 16:0/16:0 PtdIns*P* isomers. MS2 fragmentation of the PtdIns*P* (M-2H)$^{2-}$ ion at m/z 444 generates the (M-H-RCOOH)$^-$ ion at m/z 633 (following loss of 16:0 fatty acid) which, when further fragmented gives the diagnostic MS3 pattern containing the (inositol monophosphate-H$_2$O-H)$^-$ ion at m/z 241 and the m/z 409 ion (loss of 224 Da – identity unknown). The ratio of m/z 241 to m/z 409 ion intensities is diagnostic for the proportions of 16:0/16:0-PtdIns4*P* and PtdIns3*P*+ PtdIns5*P* in a sample.

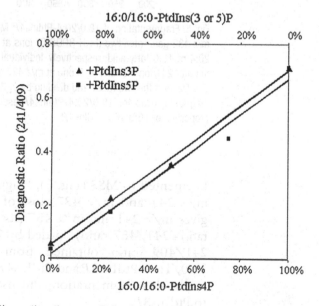

Fig. 4. Diagnostic ratio versus PtdIns*P* composition.

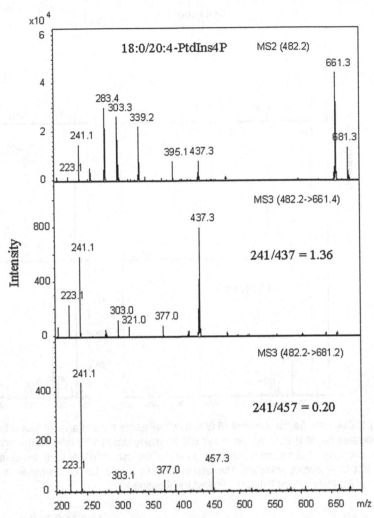

Fig. 5. Fragmentation of 18:0/20:4-PtdIns4*P*. MS2 fragmentation of the (M-2H)$^{2-}$ ion at m/z 482 generates two (M-H-RCOOH)$^{-}$ ions at m/z 661 and m/z 681 following loss of 20:4 or 18:0 fatty acid, respectively. Individual MS3 fragmentation of these generates the m/z 241 ion together with one at m/z 437 or m/z 457 respectively following loss of 224 Da. Add the two ratios and dividing by 2{(241:437 + 241:457)/2} gives a composite diagnostic ratio for 18:0/20:4-PtdIns4*P* essentially identical to that for the same proportion of 16:0/16:0-PtdIns4*P*.

fragmented to MS3 (Fig. 5). Fragmentation of m/z 661.3 gives m/z 241 and m/z 437 (loss of 224 Da), while m/z 681.3 gives m/z 241 and m/z 457 (loss of 224 Da). The (241/437 ratio + 241/457 ratio) divided by 2 is essentially identical to the 241/409 ratio obtained from the same proportions of 16:0/16:0-PtdIns3*P* and PtdIns4*P*.

In this fragmentation, PtdIns5*P* behaves almost identically to PtdIns3*P*.

3.4. Glassware Silanization

Oven dry (120°C, 60 min)and then silanize by immersing in 5% dichlorodimethylsilane in toluene (60 min, room temperature) before transferring into 100% methanol for 30 min to deactivate unreacted reagent. Remove and air dry. Store in a dark, dry atmosphere (silanization is slowly lost).

4. Notes

1. An ideal internal standard is not present in the endogenous lipid yet has a very similar structure so behaves almost identically. ^{13}C and deuterated (^{2}H) standards can be used although the former have limited availability, while the latter will undergo hydrogen exchange, resulting in deuterium loss, thus generating a range of ion masses (less obvious with freshly manufactured standards). The choice of phosphoinositide internal standards is currently very limited. The most useful are probably 16:0/16:0 (from Echelon) and 17:0/20:4 diacyl structures (from Avanti Polar Lipids); however, both are present in mammalian cells but usually at very low levels. 17:0/20:4 behaves very like 18:0/20:4, the major phosphoinositide structure in most primary mammalian cells. When working with lower eukaryotes, e.g., yeast, drosophila, which are unable to make 20:4, 17:0/20:4 is the most suitable standard.

2. PEEK (poly ether ether ketone) is probably the most inert surface available for high pressure tubing and easy to use; however, it is not fully resistant to halogenated solvents, such as chloroform, which cause it to soften and eventually burst – this is most likely to occur at tight bends. Fused silica and titanium are generally better than stainless steel but some binding of phosphoinositides and other polar lipids may still occur.

3. Extended exposure to chlorinated solvents such as chloroform in the presence of methanol and/or water will etch steel surfaces through the slow generation of HCl, thus increasing potential lipid binding sites. Re-passivation of the stainless steel with nitric acid can deactivate the surfaces again.

4. Many solvents, such as chloroform, contain stabilizers (e.g., amylene, ethanol, methanol) that can subtly alter their chromatographic characteristics. Since different manufacturers may use different stabilizers or different amounts of stabilizer, chromatographic separations may alter slightly depending on the solvent source.

5. With some analyses/instruments, postcolumn addition of a strong base like piperidine through a mixing-Tee to give a final concentration of 20 mM can improve phosphoinositide MS signal intensity. Check that the signal is improved – if not omit.

6. Most published phosphoinositide extraction methods use strongly acidic chloroform/methanol/HCl solutions; however, these protocols were found to cause extensive acid degradation, particularly of PtdInsP_2 and PtdInsP_3 (6). All plasmalogens (alkenyl structures) are also hydrolyzed. Initially, a milder citric acid/KH_2PO_4 acidification was developed, coupled with sequential butan-1-ol and chloroform/methanol extractions, which gave >90% recoveries while largely eliminating acid degradation (6); however, the procedure was relatively slow and some samples gave extensive carryover of salts and other insoluble material, which caused problems with subsequent analysis. Subsequently, a simpler, faster, cleaner procedure was developed as outlined in this chapter, but it is critical that the neutralizing EDTA wash is used prior to sample drying; else all the problems of the early methods arise.

7. If PtdInsP_3 recovery is not required then a lower HCl concentration (0.1–0.2 M) can be used to reduce the risk of acid hydrolysis.

8. Premixing the EDTA solution and methanol together will initially create a clear solution but upon standing the EDTA will start to precipitate, thus only mix these immediately prior to sample addition.

9. Approximately 4 mL synthetic lower phase is made by mixing 150 μl water, 2 mL methanolic 0.25 M HCl, 4 mL chloroform and 1.5 mL 0.25 M HCl, 2 mM sodium EDTA, 5 mM tetrabutylammonium hydrogen sulphate, and 0.9% NaCl. The mixture is allowed to separate then the upper phase is discarded leaving the synthetic lower phase.

10. While using a shallower solvent gradient can improve isomeric resolution of PtdIns$P$ and PtdInsP_2, it can sometimes lead to substantial peak broadening or even complete loss of PtdIns$(3,5)P_2$ and PtdInsP_3 for reasons that are not readily apparent. We suspect that the identical orientation of the 3 and 5 phosphates enables these structures to interact more strongly with metal and other positively charged ions than is the case with PtdIns$(4,5)P_2$ or PtdIns$(3,4)P_2$.

11. Some HPLC columns show a substantial carryover or "memory" effect, particularly for PtdInsP_2 and PtdInsP_3, with up to 10% of the signal carried into subsequent runs. This does not primarily appear to be due to poor needle or line washing but rather to binding at the head of the column. Each run cycle displaces a proportion of this material, possibly through an ion-exchange type effect, which then chromatographs as if it had been injected at the start of the run (no peak shape distortion). Repeated blank runs will remove this carryover;

however, the next sample injected on a "cleaned" column may show a reduced phosphoinositide response through re-saturation of these active sites (only when the sites are fully filled will a consistent response be seen). Making a large (10–20-fold excess) injection and chromatographic run of an unimportant phosphoinositide species immediately prior to running the real sample may overcome this problem by saturating these sites so they bind very little of the native and internal standard structures, e.g., with a mammalian sample using 17:0/20:4 as internal standard, prerun 16:0/16:0 of the same class in 20-fold excess. Carryover would be 16:0/16:0 and can be ignored.

12. Sodium is always present (it is abstracted from the solvent, column, fittings, etc.); however, the levels of sodium adduct ($(M-2H+Na)^-$, $(M-3H+2Na)^-$) formed can vary substantially depending on factors such as sample preparation technique, solvents used, sodium content of the silica, and previous samples run.

Acknowledgments

The author would like to thank Professor Michael Wakelam in whose laboratory much of this methodology was developed. This work was supported by grants from the Wellcome Trust.

References

1. Milne, S.B., Ivanova, P.T., DeCamp, D., Hsueh, R.C. and Brown, H.A. (2005) A targeted mass spectrometric analysis of phosphatidylinositol phosphate species. *J Lipid Res* **46**, 1796–1802.

2. Wenk, M.R., Lucast, L., Di Paolo, G., Romanelli, A.J., Suchy, S.F., Nussbaum, R.L., Cline, G.W., Shulman, G.I., McMurray, W. and De Camilli, P. (2003) Phosphoinositide profiling in complex lipid mixtures using electrospray ionization mass spectrometry. *Nat Biotechnol* **21**, 813–817.

3. Michelsen, P., Jergil, B. and Odham, G. (1995) Quantification of phosphoinositides using selected-ion monitoring electrospray mass-spectrometry. *Rapid Commun Mass Spectrom* **9**, 1109–1114.

4. Hsu, F.-F. and Turk, J. (2000) Characterisation of phosphatidylinositol, phosphatidylinositol-4-phosphate and phosphatidylinositol-4,

5-bisphosphate by electrospray ionizationtandem mass spectrometry: a mechanistic study. *Am Soc Mass Spectrom* **11**, 986–999.

5. Pettitt, T.R. (2008) Lipidomic analysis of phospholipids and related structures by liquid chromatography-mass spectrometry. In *Methods in Molecular Biology* (Larijani, B., Woscholski, R. & Rosser, C.A., eds), Humana, Totowa, NJ. Vol **462**, 25–41.

6. Pettitt, T.R., Dove, S.K., Lubben, A., Calaminus, S.D.J. and Wakelam, M.J.O. (2006) Analysis of intact phosphoinositides in biological samples. *J Lipid Res* **47**, 1588–1596.

7. Tuytten, R., Lemiere, F., Witters, E., Van Dongen, W., Slegers, H., Newton, R.P., Van Onckelen, H. and Esmans, E.L. (2006) Stainless steel electrospray probe: a dead end for phosphorylated organic compounds? *J Chromatogr A* **1104**, 209–221.

Chapter 14

Imaging Phosphoinositide Dynamics in Living Cells

Anne Wuttke, Olof Idevall-Hagren, and Anders Tengholm

Abstract

To improve our understanding of the important roles played by inositol lipid derivatives in signalling and other cellular processes, it is crucial to measure phosphoinositide concentration changes in individual cells with high spatial and temporal resolution. A number of protein domains that interact with inositol lipids in a specific manner have been identified. Tagged with the green fluorescent protein or its colour variants, these protein modules can be used as probes to visualize various phosphoinositide species in different sub-cellular compartments. Here, we present protocols for fluorescence imaging of phosphoinositide dynamics in single living cells. Total internal reflection fluorescence microscopy is particularly powerful for time-lapse recordings of phosphoinositides in the plasma membrane. We demonstrate how this technique can be used to record phospholipase C- and PI3-kinase-induced changes in inositol lipids in insulin-secreting cells. These procedures should be applicable to studies of the spatio-temporal regulation of phosphoinositide metabolism in many types of cells.

Key words: Phosphatidylinositol 4,5-bisphosphate, Phosphatidylinositol 3,4,5-trisphosphate, Phospholipase C, PI3-kinase, Pleckstrin homology domain, Ca^{2+}, Green fluorescent protein, Total internal reflection fluorescence microscopy, Insulin-secreting cell

1. Introduction

Given the crucial function of phosphoinositides (PIs) and inositol phosphates in many biological processes, it is essential to improve the understanding of their spatio-temporal regulation in various cells and tissues. Biochemical techniques based on radiotracer labelling, chromatography or mass spectrometry are of great value for investigating PI profiles in cell and tissue extracts. They yield quantitative information and can be optimized to distinguish between all different PI isomers (1). However, these methods have also important limitations; the major ones being that large numbers of cells are required to obtain a sufficient signal and that

Christopher J. Barker (ed.), *Inositol Phosphates and Lipids: Methods and Protocols*, Methods in Molecular Biology, vol. 645, DOI 10.1007/978-1-60327-175-2_14, © Humana press, a part of Springer Science+Business Media, LLC 2010

repeated measurements cannot be made from the same sample. These shortcomings preclude detection of compartmentalized signals, short-lasting transients, oscillations, and other complex time-courses that characterize many signalling systems.

Over the past decade the development of genetically encoded fluorescent biosensors for single cell detection of PIs has greatly improved our understanding of PI lipid signalling. Pioneered by the Meyer (2) and Balla (3) laboratories, the use of green fluorescent protein (GFP) or its spectral variants fused to protein modules with specific PI-binding properties has become the most valuable tool in the investigation of PIs in single living cells. Analysis of fluorescent reporter protein localization and stimulus-induced translocation provides information about the intracellular distribution and changes in relative levels of a particular lipid.

A large number of protein domains have been identified that are useful for following changes in a wide variety of intracellular lipids (4–7). A non-exhaustive list of such protein domains for detection of particular PI lipids in single cells is presented in Table 1. Fab1p-YOTB-Vac1p-EEA1 (FYVE) domains and phox homology (PX) domains bind phosphatidylinositol-3-phosphate and perhaps other 3'-phosphorylated inositides. The pleckstrin homology (PH) domains constitute the largest family of PI-binding protein modules and are the most commonly employed probes in imaging studies. The availability of fairly extensive information on in vitro binding specificity probably contributes to their popularity in imaging applications. Most PH domains bind membrane PIs with low affinity and little specificity (5), but there are PH domains with relatively high affinity and specificity for most of the 4'- and 5'-phosphorylated PI lipids.

The PH domain from phospholipase C-δ1 (PLCδ1$_{PH}$) is the best characterized PI-binding protein domain. It binds strongly in vitro to phosphatidylinositol-4,5-bisphosphate (PtdIns(4,5)P_2) and with even higher affinity to its isolated headgroup D-myo-inositol-1,4,5-trisphosphate (InsP_3) (8), and is therefore a valuable tool for assessing PtdIns(4,5)P_2 localization and phospholipase C (PLC) activity in individual cells. In unstimulated cells, the GFP-tagged PLCδ1$_{PH}$ is bound to the plasma membrane, where the concentration of PtdIns(4,5)P_2 is relatively high. Upon activation of PLC, which hydrolyses PtdIns(4,5)P_2 into InsP_3 and diacylglycerol, the biosensor dissociates from the membrane and binds InsP_3 in the cytoplasm (Fig. 1a). Additional examples of PH domains with high PI-binding specificity include those from protein kinase B/Akt and GRP1 (general receptor for phosphoinositides-1), which have been extensively used for detecting phosphatidylinositol-3,4-bisphosphate (PtdIns(3,4)P_2) and phosphatidylinositol-3,4,5-trisphosphate (PtdIns(3,4,5)P_3) (9–11), the lipid products of phosphoinositide-3-OH-kinase (PI3-kinase) activity. As the concentration of PtdIns(3,4)P_2 and PtdIns(3,4,5)P_3

Table 1
Protein modules used with fluorescent tags for visualization of specific phospho-inositides in single cells.

Lipid	Protein module	Reference	Comment
PtdIns3P	EEA1-(2x)-FYVE	(22–25)	EEA1-GFP also binds phosphatidylserine and phosphatidylinositol in vitro. The probe interferes with early endosome formation.
	Hrs-(2x)-FYVE	(22, 24, 26)	
	P40^{phox}-PX	(27–30)	
PtdIns4P	OSBP-PH	(31–33)	Binding to the Golgi apparatus is determined also by PtdIns4P-independent mechanisms; binds also PtdIns (4,5) P_2 in vitro.
	FAPP1-PH	(32–34)	See comment for OSBP-PH above.
	OSH2-PH(2x)	(35, 36)	Detects PtdIns4P in the plasma membrane but not in the Golgi compartment in mammalian cells.
	OSH1-PH	(35, 36)	Localizes to the Golgi apparatus and plasma membrane.
PtdIns5P	ING2-PHD	(37)	Some binding to PtdIns3P and PtdIns4P in vitro, but no colocalization with p40^{phox}-PX in vivo.
PtdIns(3,4)P_2	p47^{phox}-PX	(27, 38, 39)	Specificity has been questioned (39). A portion of the domain binds anionic phospholipids, such as phosphatidic acid and phosphatidylserine; intramolecular association with the C-terminal SH3 domain might mask membrane interaction.
	TAPP1-PH	(40–42)	Membrane penetration slows dissociation. A mutant exists with less penetration and faster dissociation kinetics (41).
PtdIns(4,5)P_2	PLCδ1-PH	(2, 3, 43)	PLCδ1 contains the best characterized PH domain, which binds strongly to both PtdIns(4,5)P_2 and its isolated headgroup, D-*myo*-inositol-1,4,5-trisphosphate. Does not readily recognize PtdIns(4,5)P_2 in membranes other than the plasma membrane.
	Sla2-ANTH	(44)	Used to detect PtdIns(4,5)P_2 at sites of endocytosis.
PtdIns(3,5)P_2	Atg18	(40, 45–47)	May also bind PtdIns3P. PH domains specific for PtdIns(3,5)P_2 in vitro have been identified (Centaurin-β2 and α-Syntrophin), but not yet utilized as molecular probes.

(continued)

Table 1
(continued)

Lipid	Protein module	Reference	Comment
$PtdIns(3,4,5)P_3$	Akt-PH	(10, 11, 41)	Binds also to $PtdIns(3,4)P_2$. Has been used in a FRET-based biosensor (21).
	GRP1-PH	(10, 41)	Binds to $PtdIns(3,4)P_2$ with much lower affinity. Has been used in a FRET-based biosensor (20)
	ARNO-PH	(41, 48)	Low affinity for $PtdIns(3,4,5)P_3$. Binds also $PtdIns(3,4)P_2$.
	Btk-PH	(41, 49)	Specific for $PtdIns(3,4,5)P_3$.

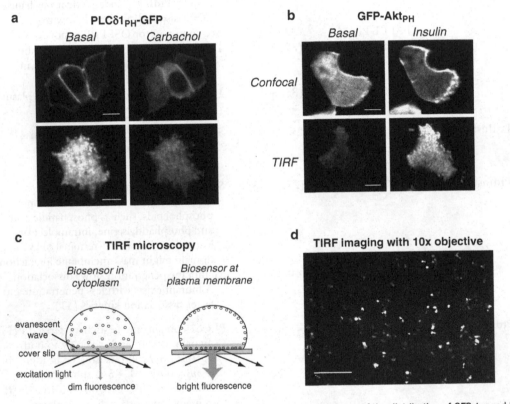

Fig. 1. Confocal and total internal reflection fluorescence (TIRF) microscopy images of the distribution of GFP-tagged PH domains in insulin-secreting MIN6-cells. (**a**) The $PtdIns(4,5)P_2/InsP_3$-binding PH domain from PLCδ1 (PLCδ1$_{PH}$-GFP) shows strong localization to the plasma membrane under basal conditions, but translocates to the cytoplasm after activation of phospholipase C with 100 μM of the muscarinic agonist carbachol. The loss of GFP from the membrane is seen as decreased fluorescence in the TIRF image. (**b**) The GFP-tagged $PtdIns(3,4)P_2/PtdIns(3,4,5)P_3$-binding PH domain from Akt (GFP-Akt$_{PH}$) shows cytoplasmic and nuclear distribution in unstimulated cells, but translocates to the plasma membrane after activation of PI3-kinase with 100 nM insulin. The accumulation of GFP at the plasma membrane is seen as increased fluorescence in the TIRF image. Scale bars, 5 μm. (**c**) The principle of TIRF microscopy. An exponentially decaying electromagnetic field ("evanescent wave") is generated by the total internal reflection of the laser excitation light at the top surface of the coverslip. This field excites fluorescent molecules within ~100 nm above the glass, corresponding to the plasma membrane and immediate sub-membrane space. The fluorescence intensity correlates with the degree of membrane localization of the fluorescent protein construct. (**d**) Large area evanescent wave illumination of MIN6 cells expressing GFP-tagged GRP1 with fluorescence imaging through a 10× objective. Scale bar, 200 μm.

is very low in unstimulated cells, fluorescent protein-tagged versions of these PH domains are distributed in the cytoplasm and nucleus. Receptor stimuli which activate PI3-kinase cause translocation of the biosensors to the plasma membrane (Fig. 1b). For more detailed information on the different PI-binding protein modules, the reader is referred to the references in Table 1 and to recent comprehensive reviews (4–7, 12, 13).

The localization of the fluorescent biosensors is most commonly analysed with confocal microscopy, which provides high resolution images of optical sections of the specimen. This technique is well suited for visualization of different sub-cellular compartments such as the plasma membrane, cytoplasm, and intracellular organelles. For cell types that are very flat, regular wide-field epifluorescence microscopy might work equally well. In many cases, the interesting changes in PI lipids take place in the plasma membrane. A technique known as evanescent wave microscopy or total internal reflection fluorescence (TIRF) microscopy provides selective imaging of fluorescence in a thin volume of the cell where the plasma membrane is in contact with the coverslip (14, 15). In this method a laser beam is directed towards a coverslip with adhering cells at an angle causing total reflection at the interface between the coverslip and the lower refractive index aqueous medium surrounding the cells. Excitation is provided by an evanescent wave travelling in a zone within ~100 nm above the glass-water interface, exciting fluorescent molecules exclusively in the plasma membrane and the immediate sub-membrane space (Fig. 1c). Studies in the authors' laboratory have employed TIRF microscopy for measurements of the plasma membrane $PtdIns(4,5)P_2$ and $PtdIns(3,4,5)P_3$ concentrations using the PH domains from PLCδ1 and protein kinase B/Akt (16–19). TIRF microscopy provides better resolution as the optical section is several-fold thinner than in confocal microscopy. Since most of the sample is not illuminated, there is little background, which provides better signal-to-noise ratio and thereby increased sensitivity to detect small fluorescence changes. Moreover, photobleaching and phototoxicity are greatly reduced, which permits measurements over longer periods of time and with better time resolution without damaging the specimen. Another major advantage of TIRF microscopy is that the detection of biosensor translocation to or from the plasma membrane only requires simple fluorescence intensity recordings, which can be made on many cells in parallel (Fig. 1d).

Several excellent reviews have been published on single cell imaging of phosphoinositides and advantages and limitations of this approach (4, 6, 13). Here, we detail the procedures used in our laboratory for time-lapse image analysis of PI levels in the plasma membrane of individual living cells. Although particular emphasis is made on TIRF microscopy recordings of phospholipase C- and PI3-kinase-induced changes in $PtdIns(4,5)P_2$ and

PtdIns$(3,4,5)P_3$ in insulin-secreting cells, the protocols should be generally applicable to other cell types and signalling pathways. In many cases, it is important to study the relationship between the changes in PI lipids and other signalling events. As an example, we have included a description of parallel recordings of lipid signals and the cytoplasmic Ca^{2+} concentration.

2. Materials

2.1. Cell Culture and Transfection

1. Glass coverslips, 25-mm diameter, 0.08–0.13 mm thickness (from e.g. Menzel-Gläser, Germany).
2. Poly-L-lysine (MW 70,000–150,000; Sigma, St Louis, MO): Prepare 0.1 mg/mL stock solution in sterile water and store at –20°C. Prepare the working solution at a concentration of 0.01 mg/mL in sterile water.
3. Dulbecco's phosphate buffered saline (D-PBS; Invitrogen, Carlsbad, CA): 137 mM NaCl, 2.67 mM KCl, 8.10 mM Na_2HPO_4, 1.47 mM KH_2PO_4, store at +4°C.
4. Complete growth medium for MIN6 β-cells is prepared from Dulbecco's Modified Eagle's Medium (DMEM) containing 4,500 mg/L glucose (Invitrogen) by supplementing with 15% (v/v) fetal calf serum, 2 mM L-glutamine, 70 µM β-mercaptoethanol, 100 U/mL penicillin, and 100 µg/mL streptomycin. Store at +4°C.
5. For transient transfection of MIN6 cells, use Lipofectamine™ 2000 reagent (Invitrogen) and OptiMEM I (Invitrogen). The Lipofectamine™ 2000 reagent should be mixed gently before use. Store at +4°C.
6. Plasmid DNA encoding the PI biosensor of interest.
7. Fura Red acetoxymethyl ester (Molecular Probes Invitrogen) for measurements of the cytoplasmic Ca^{2+} concentration.

2.2. Fluorescence Microscopy

1. Experimental buffer for incubation and superfusion: 125 mM NaCl, 4.9 mM KCl, 1.3 mM $CaCl_2$, 1.2 mM $MgCl_2$, 25 mM HEPES, 3 mM D-glucose, and 0.1% (w/v) bovine serum albumin. Adjust pH with NaOH to 7.40 at 37°C. This buffer is made fresh on the day of experiment.
2. Confocal or TIRF fluorescence microscope: While a regular fluorescence microscope may occasionally work for some very flat cell types, a microscope with optical sectioning capability is advantageous in the majority of cases. Confocal microscopy is required for analysis of PI lipids in intracellular organelles, whereas TIRF is better for imaging the plasma membrane.

(a) Many suitable confocal systems are available through the major microscope manufacturers. The authors use a Yokogawa CSU-10 spinning disk system (available from several distributors in Europe and the US), well suited for live cell imaging in permitting a high acquisition speed and yet relatively low degree of photobleaching.

(b) TIRF illumination can easily be obtained through the objective lens using a high NA (>1.40) objective and an illuminator device for beam focussing and positioning (available from many of the major microscope manufacturers; illuminator also available from TILL Photonics/ Agilent Technology, Santa Clara, CA). An alternative TIRF configuration uses a prism for illumination, which has the advantage of allowing imaging with a low magnification objective to obtain information from many cells in parallel (Fig. 1d). The authors use a custom-built system, but commercial illuminators are available from, e.g. TIRF Technologies (Morrisville, NC).

3. Light source, filters for wavelength selection, and equipment for image acquisition and analysis: Commercial confocal microscope systems are usually equipped with all required peripheral instrumentation. A TIRF system often requires custom configuration.

(a) The argon ion laser is a common light source for excitation of fluorescent proteins. GFP is excited with the 488-nm line and the 458-nm and 514-nm lines can be used for excitation of cyan and yellow fluorescent protein and even of some red fluorescent proteins. Solid-state lasers are becoming increasingly popular and are available in a variety of spectral variants well suited for imaging GFP and related fluorescent proteins.

(b) Appropriate excitation filter (e.g. 488 nm/10 nm half-bandwidth), dichroic mirror (e.g. 500DCXRU), and emission filters (e.g. 525/40 nm for GFP and 630 nm long-pass for Fura Red) can be obtained from Chroma Technology (Rockingham, VT), Omega Optical (Battleboro, VT), or Semrock (Rochester, NY). Dual channel recordings may require a filter changer, which can be obtained from, e.g. Sutter Instruments (Novato, CA).

(c) A sensitive low-light level charge-coupled device (CCD) camera for spinning disk confocal or TIRF microscope can be obtained from, e.g., Andor Technology, Hamamatsu, Princeton Instruments, and Cooke corporation.

(d) Data acquisition and analysis software are usually included with commercial confocal systems but may have to be purchased separately for custom-built setups. The authors use MetaFluor from Molecular Devices (Downington, PA) and the free ImageJ software (Rasband, W.S., National Institutes of Health, Bethesda, MD, http://rsb.info.nih.gov/ij/).

4. Equipment for buffer superfusion and thermoregulation: It is often important to be able to add and remove test substances as well as to maintain a temperature of 37°C during the experiment. We use a peristaltic pump in combination with a custom-built superfusion chamber, chamber heater, and microscope objective heater. Similar equipment is available from, e.g., Warner Instruments (Hamden, CT).

3. Methods

3.1. Preparation of Poly-L-lysine Coated Coverslips

1. Wash the coverslips in deionized water and sterilize in a heat incubator (180°C) or by dipping them into 95% ethanol and passing through the flame of a Bunsen burner. Place each sterile coverslip in a 35-mm diameter culture dish.

2. Place a ~100 µL drop of poly-L-lysine (0.01 mg/mL) on the coverslip, leave it for 20–30 min at room temperature, and then remove the poly-L-lysine and rinse the coverslip twice with sterile water. It is important not to let the poly-L-lysine dry on the coverslips. After the last washing step, remove as much water as possible and allow the coverslips to dry in a laminar flow bench with the lids of the culture dishes open (see Note 1).

3.2. Cell Culture and Transient Transfection

1. Seed cells onto the dried poly-L-lysine coated coverslips by adding 3 mL cell suspension at a concentration of 0.1–0.2×10^6/mL. The confluency should be approximately 50% on the day of transfection (see Note 2).

2. For each coverslip to be transfected, mix 2 µg of plasmid DNA encoding the PI biosensor of interest with 125 µL OptiMEM I in one tube and 5 µL Lipofectamine™ 2000 with 125 µL OptiMEM I in another tube. Leave the tubes for 5 min at room temperature (see Note 3).

3. Combine the content of the two tubes and mix gently (do not vortex). Leave for 20 min at room temperature to allow DNA-liposome complexes to form.

4. Remove complete culture medium from the cells and carefully rinse them once with D-PBS preheated to 37°C.

5. Remove the D-PBS and add to each coverslip 750 µL OptiMEM I conditioned to 37°C. Subsequently add the

DNA-liposome combination (~250 μL per coverslip) and mix by gently rocking the culture dish back and forth.

6. Incubate the cells for 3–5 h at 37°C and 5% CO_2 in a cell culture incubator.

7. Terminate the transfection reaction by removing the transfection medium, rinsing the cells once with D-PBS at 37°C, and adding 3 mL complete growth medium.

8. Culture the transfected cells for 12–48 h before imaging experiments to allow expression of the PI biosensor (see Note 4).

3.3. Time-Lapse Imaging of Phosphoinositide Dynamics in Living Cells

1. Rinse the cells on the coverslip once with the experimental buffer, add 2 mL new buffer to the dish and incubate for 30–60 min in a 37°C incubator to allow cells to adapt to basal conditions in the experimental buffer (see Note 5).

2. Place the coverslip with cells in a suitable superfusion chamber that fits the stage of the microscope.

3. Place the chamber on the thermostated microscope stage, connect the chamber to a peristaltic pump, and superfuse cells with experimental buffer at a rate of 0.5–1 mL/min. The speed should be such that the entire chamber volume is exchanged several times per minute. Make sure that the temperature is 37°C in the chamber (see Note 6).

4. Observe the transfected cells in the fluorescence microscope with appropriate settings for fluorescent protein excitation and emission and select a region of the coverslip with cells expressing moderate levels of the PI biosensor (see Note 7).

5. Project the image on the CCD camera and adjust acquisition parameters, such as exposure time and signal gain to appropriate levels (see Note 8).

6. Move stage to find a region on the coverslip without cells. Acquire a background image without changing imaging parameters. Return to the previously identified coverslip region with the cells.

7. Acquire images with the CCD camera every 1–5 s.

8. Analyse changes of relative fluorescence in individual cells. Most software permits plotting of intensity changes on-line to help monitoring the experiment, but in addition a more careful off-line analysis is required. Subtract the background image and plot fluorescence from regions of interest using suitable software (e.g. MetaMorph or MetaFluor from Molecular Devices, or ImageJ, W.S.Rasband, http://rsb.info.nih.gov/ij). Different analysis strategies are required depending on the type of images (see Fig. 2 and Note 9).

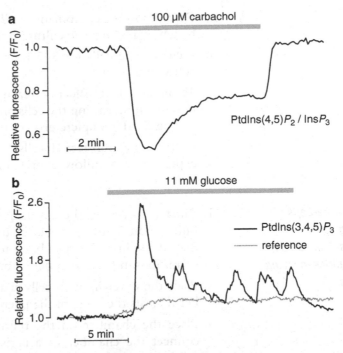

Fig. 2. Single and dual wavelength time-lapse TIRF microscopy recordings of PI biosensor membrane localization in individual MIN6 cells. (a) Activation of phospholipase C with the muscarinic receptor agonist carbachol results in dissociation of PLCδ1$_{PH}$-GFP from the plasma membrane with concomitant loss of fluorescence that is reversed when the stimulus is removed. The biphasic response with a strong initial response followed by a sustained plateau has been found to result from distinct mechanisms of Ca^{2+} feedback on phospholipase C (17). (b) Elevation of the glucose concentration from 3 to 11 mM triggers oscillations in the membrane concentration of PtdIns(3,4,5)P_3 as recorded with GFP-tagged GRP1 (*black trace*). These changes are not due to alterations of cell shape or adhesion, since simultaneous measurements of the red fluorescent protein tdimer2 (50) anchored to the membrane with a polybasic sequence and a C-terminal prenylation motif (51) show a stable signal (*grey trace*). The glucose-induced PtdIns(3,4,5)P_3 oscillations have been found to reflect pulsatile release of insulin, resulting in autocrine stimulation of insulin receptors and intermittent activation of PI3-kinase (19). GFP-GRP1 and membrane-anchored tdimer2 were co-expressed in the same cell. GFP was excited at 488 nm and tdimer2 at 514 nm using an argon ion laser. Emission was measured at 525/40 nm for GFP and 630 nm long pass for tdimer2.

3.4. Simultaneous Recordings of Cytoplasmic Ca^{2+} Concentration and Phosphoinositide Dynamics in Living Cells

1. Rinse the transfected cells twice with experimental buffer, add 2 mL of the buffer supplemented with 10 µM of the acetoxymethyl ester of the Ca^{2+} indicator Fura Red, and incubate for 45 min at 37°C and protect from light (see Note 10).

2. Rinse the cells thoroughly with experimental buffer to remove extracellular indicator.

3. Mount the coverslip in the microscope superfusion chamber and record fluorescence as described in Subheading 3.3 (see Fig. 3 and Note 11).

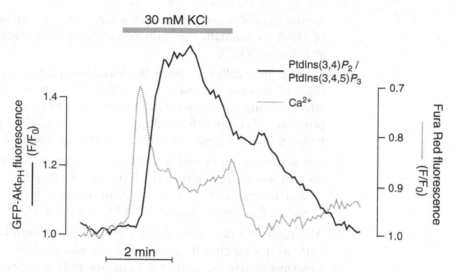

Fig. 3. Simultaneous TIRF microscopy recordings of membrane PtdIns(3,4)P_2/PtdIns(3,4,5)P_3 concentration with GFP-Akt$_{PH}$ and the sub-membrane cytoplasmic Ca^{2+} concentration with Fura Red in a single insulin-secreting MIN6 cell. Depolarization of the plasma membrane by raising the KCl concentration to 30 mM results in voltage-dependent Ca^{2+} influx and elevation of the cytoplasmic Ca^{2+} concentration. Increase of Ca^{2+} is associated with loss of Fura Red fluorescence (*grey trace*), but the trace has been inverted to show the increase in Ca^{2+} as an upward deflection of the curve. It is apparent that the increase in Ca^{2+} precedes that of GFP-Akt$_{PH}$ fluorescence, which is due to Ca^{2+} triggering the exocytosis of insulin, which subsequently activates insulin receptors and PI3-kinase via autocrine feedback.

4. Notes

1. As an alternative to poly-L-lysine, coverslips can be coated with collagen, fibronectin, or gelatine. For gelatine coating, add 1% (w/v) sterile gelatine in D-PBS to the coverslips, incubate for 1 h in a cell culture incubator at 37°C in a humidified atmosphere, remove the gelatine, and wash twice with D-PBS. Seed cells onto the coverslips immediately, because the gelatine must not dry.

2. This protocol typically yields 30–40% transfection efficiency 1–2 days after transfection. The best results are obtained if the cells are allowed to grow on the coverslip for 36–48 h prior to transfection. The protocol is optimized for insulin-secreting MIN6 and INS-1 cells and may have to be modified for other cell lines. Obviously, other methods for transfection can be used. For example, adenovirus-mediated gene transfer is highly efficient in a large variety of cell types.

3. If the experiment requires transfection of more than one plasmid, all cDNA should be transfected at the same time to obtain maximal cotransfection. It is not recommended to use more than a total of 3 μg plasmid DNA per coverslip as higher concentrations may have adverse effects on the cells and contribute to reduction of the transfection efficiency.

Adjust the volume of liposomes when changing the amount of DNA to maintain the ratio of DNA:Lipofectamine 2000 at 1:2.5 (w/v).

4. Do not allow cells to express the biosensor protein for more than 48 h, since excessive levels of the fusion protein may affect cell function. This is because the PI-binding fluorescent proteins will compete for endogenous effectors, which potentially inhibits the lipid-mediated cellular responses (13).

5. For some cell types and applications longer pre-incubation time in serum-free culture medium may be required (up to several hours), since serum contains factors that may e.g. stimulate PI3-kinase formation of $PtdIns(3,4,5)P_3$.

6. Although cells can be stimulated by adding medium into the bath with a pipette, it is preferable to add medium using a superfusion system. Such a system not only permits convenient washout of the stimulus but also eliminates problems with evaporation that otherwise would occur, since the ambient temperature should be maintained at 37°C to observe normal biosensor localization and translocation responses.

7. There will be a large variability between cells in the expression level of the fluorescent biosensor. High levels of phosphoinositide binding proteins can interfere with processes in the cells as mentioned in Note 4. Choose cells with relatively low expression levels, but not so low that the signal-to-noise ratio is compromised. The fluorescence collected from the cells expressing the biosensor should be well above the autofluorescence from non-transfected cells. Also avoid cells with large fluorescent aggregates, as this often reflects accumulation of improperly folded proteins in the endoplasmic reticulum or Golgi apparatus.

8. Select camera exposure time and gain settings so that no pixels in the image will be saturated. It is important to consider that the fluorescence intensity might change several-fold during an experiment if the biosensor redistributes between different cellular compartments. It is good to keep the exposure times as short as possible without compromising signal-to-noise ratio, since excessive exposure to excitation light may result in photobleaching and phototoxic effects. If the fluorescence collected from the cells is very low, it is possible with most CCD cameras to combine charges in adjacent pixels to form one pixel in a process named binning. This will enhance the signal at the expense of optical resolution. On the contrary, if the signal is too bright, it indicates an excessive excitation light intensity. Reduce the laser power or attenuate the light with neutral density filters in the excitation beam path. The laser beam should be completely blocked with a shutter

between image captures to avoid adverse effects of the light on the specimen.

9. (a) For confocal microscopy images, the simplest way to assess plasma membrane translocation and dissociation of the biosensor is to quantify the changes in cytoplasmic fluorescence by defining a region of interest inside the cell that excludes the nucleus, membrane or other conspicuous organelles. Alternatively, regions of interest can be placed over the plasma membrane either manually or with the help of a segmentation algorithm. This is often a difficult method because the cell shape tends to change over the time-course of an experiment, in particular after cell stimulation.

(b) In TIRF microscopy images, a region of interest over the cell will always show fluorescence in the plasma membrane and translocation or dissociation is simply recorded as changes of intensity.

(c) Intensity changes can easily be expressed in relation to baseline by dividing the intensity value at each time point with the prestimulatory level (F/F_0).

(d) Some cell types exhibit profound membrane ruffling or other morphological changes, which can alter the area of membrane contact with the coverslip, especially after stimulation. In a TIRF microscope such morphological changes can induce a fluorescence change that might be misinterpreted as translocation. It is therefore important to perform control experiments to ensure that an observed fluorescence change is indeed due to biosensor translocation. The translocation response can be confirmed with confocal microscopy. An alternative and more powerful approach is to use a reference fluorescent protein that remains soluble in the cytoplasm or attached to the plasma membrane. Ideally, such an experiment should be performed by co-expressing the PI biosensor tagged with one fluorescent protein and the reference protein in a different colour to enable simultaneous recordings of the two constructs in the same cell (Fig. 2).

(e) As an alternative to translocation readout, some PI-binding protein domains have been used in biosensors that rely on fluorescence resonance energy transfer between cyan and yellow fluorescent proteins as a result of a conformational change in the protein domain upon lipid binding (20, 21). However, proper detection of fluorescence resonance energy transfer is technically difficult and the dynamic range of the signals is typically much smaller than for the translocation-based biosensors.

(f) It is important to keep in mind that a certain PI-binding protein domain often does not recognize all pools of the particular lipid because of competition with endogenous effectors binding to the lipid. For example, the $PLC\delta1_{PH}$ recognizes $PtdIns(4,5)P_2$ exclusively in the plasma membrane, although the lipid is known to occur also in intracellular membranes. In addition, the localization of the PI-binding protein domains is not only determined by binding to lipids, but may also be affected by interactions with other proteins. See ref. 13 for a more thorough discussion on these limitations.

10. Dissolve Fura Red at 10 mM in DMSO and store at −20°C protected from light. Mix before use and incubate together with cells at 2–20 µM during 20–45 min. The optimal indicator concentration and incubation time vary depending on cell type and cell density. Excessive loading may reduce the Ca^{2+} response by buffering the concentration changes, whereas too little loading gives poor signal-to-noise ratio.

11. Fura Red is readily excited by the same light source as GFP, such as the 488-nm line of the argon ion laser. Fluorescence emission is detected with a 630 nm long-pass filter. Upon Ca^{2+} binding, the excitation spectrum of Fura Red becomes blue-shifted resulting in loss of 488 nm excited fluorescence. With appropriate emission filters, the fluorescence from Fura Red can be well separated from that of GFP-labelled PI biosensors. Alternation between the Fura Red and GFP emission filters using a filter wheel or similar device allows simultaneous measurements of cytoplasmic Ca^{2+} concentration and PI lipid dynamics in the same cell. However, as GFP shows some emission even above 600 nm, there may be a slight cross-over of GFP fluorescence into the Fura Red channel in cells with very bright GFP fluorescence and poor Fura Red loading. The degree of overlap is easily estimated by imaging cells expressing GFP, but lacking the Ca^{2+} indicator, using the Fura Red filter set.

Acknowledgments

We thank Professors Tobias Meyer, Stanford University, for the GFP-PH_{Akt} and $PLC\delta1_{PH}$-GFP plasmids and Roger Tsien, University of California in San Diego, for tdimer2. The authors' work is supported by grants from Åke Wiberg's Foundation, the European Foundation for the Study of Diabetes/MSD, the family Ernfors Foundation, Harald and Greta Jeanssons Foundations, Novo Nordisk Foundation, the Swedish Diabetes Association and the Swedish Research Council.

References

1. Rusten, T. E., and Stenmark, H. (2006) Analyzing phosphoinositides and their interacting proteins. *Nat Methods* **3**, 251–258.

2. Stauffer, T. P., Ahn, S., and Meyer, T. (1998) Receptor-induced transient reduction in plasma membrane PtdIns(4,5)P_2 concentration monitored in living cells. *Curr Biol* **8**, 343–346.

3. Varnai, P., and Balla, T. (1998) Visualization of phosphoinositides that bind pleckstrin homology domains: calcium- and agonist-induced dynamic changes and relationship to myo-[^{3}H]inositol-labeled phosphoinositide pools. *J Cell Biol* **143**, 501–510.

4. Halet, G. (2005) Imaging phosphoinositide dynamics using GFP-tagged protein domains. *Biol Cell* **97**, 501–518.

5. Lemmon, M. A. (2008) Membrane recognition by phospholipid-binding domains. *Nat Rev Mol Cell Biol* **9**, 99–111.

6. Balla, T., and Varnai, P. (2002) Visualizing cellular phosphoinositide pools with GFP-fused protein-modules. *Sci STKE* **2002**, PL3.

7. Lemmon, M. A. (2003) Phosphoinositide recognition domains. *Traffic* **4**, 201–213.

8. Lemmon, M. A., Ferguson, K. M., O'Brien, R., Sigler, P. B., and Schlessinger, J. (1995) Specific and high-affinity binding of inositol phosphates to an isolated pleckstrin homology domain. *Proc Natl Acad Sci USA* **92**, 10472–10476.

9. Haugh, J. M., Codazzi, F., Teruel, M., and Meyer, T. (2000) Spatial sensing in fibroblasts mediated by 3′ phosphoinositides. *J Cell Biol* **151**, 1269–1280.

10. Gray, A., Van Der Kaay, J., and Downes, C. P. (1999) The pleckstrin homology domains of protein kinase B and GRP1 (general receptor for phosphoinositides-1) are sensitive and selective probes for the cellular detection of phosphatidylinositol 3,4-bisphosphate and/or phosphatidylinositol 3,4,5-trisphosphate in vivo. *Biochem J* **344 Pt 3**, 929–936.

11. Tengholm, A., and Meyer, T. (2002) A PI3-kinase signaling code for insulin-triggered insertion of glucose transporters into the plasma membrane. *Curr Biol* **12**, 1871–1876.

12. Downes, C. P., Gray, A., and Lucocq, J. M. (2005) Probing phosphoinositide functions in signaling and membrane trafficking. *Trends Cell Biol* **15**, 259–268.

13. Varnai, P., and Balla, T. (2006) Live cell imaging of phosphoinositide dynamics with fluorescent protein domains. *Biochim Biophys Acta* **1761**, 957–967.

14. Axelrod, D. (2001) Total internal reflection fluorescence microscopy in cell biology. *Traffic* **2**, 764–774.

15. Steyer, J. A., and Almers, W. (2001) A real-time view of life within 100 nm of the plasma membrane. *Nat Rev Mol Cell Biol* **2**, 268–275.

16. Thore, S., Dyachok, O., and Tengholm, A. (2004) Oscillations of phospholipase C activity triggered by depolarization and Ca^{2+} influx in insulin-secreting cells. *J Biol Chem* **279**, 19396–19400.

17. Thore, S., Dyachok, O., Gylfe, E., and Tengholm, A. (2005) Feedback activation of phospholipase C via intracellular mobilization and store-operated influx of Ca^{2+} in insulin-secreting β-cells. *J Cell Sci* **118**, 4463–4471.

18. Thore, S., Wuttke, A., and Tengholm, A. (2007) Rapid turnover of phosphatidylinositol-4,5-bisphosphate in insulin-secreting cells mediated by Ca^{2+} and the ATP-to-ADP ratio. *Diabetes* **56**, 818–826.

19. Idevall-Hagren, O., and Tengholm, A. (2006) Glucose and insulin synergistically activate PI3-kinase to trigger oscillations of phosphatidylinositol-3,4,5-trisphosphate in beta-cells. *J Biol Chem* **281**, 39121–39127.

20. Sato, M., Ueda, Y., Takagi, T., and Umezawa, Y. (2003) Production of PtdInsP$_3$ at endomembranes is triggered by receptor endocytosis. *Nat Cell Biol* **5**, 1016–1022.

21. Ananthanarayanan, B., Ni, Q., and Zhang, J. (2005) Signal propagation from membrane messengers to nuclear effectors revealed by reporters of phosphoinositide dynamics and Akt activity. *Proc Natl Acad Sci USA* **102**, 15081–15086.

22. Gillooly, D. J., Morrow, I. C., Lindsay, M., Gould, R., Bryant, N. J., Gaullier, J. M., Parton, R. G., and Stenmark, H. (2000) Localization of phosphatidylinositol 3-phosphate in yeast and mammalian cells. *EMBO J* **19**, 4577–4588.

23. Burd, C. G., and Emr, S. D. (1998) Phosphatidylinositol(3)-phosphate signaling mediated by specific binding to RING FYVE domains. *Mol Cell* **2**, 157–162.

24. Gaullier, J. M., Simonsen, A., D'Arrigo, A., Bremnes, B., Stenmark, H., and Aasland, R. (1998) FYVE fingers bind PtdIns(3)P. *Nature* **394**, 432–433.

25. Patki, V., Lawe, D. C., Corvera, S., Virbasius, J. V., and Chawla, A. (1998) A functional PtdIns(3)P-binding motif. *Nature* **394**, 433–434.

26. Vermeer, J. E., van Leeuwen, W., Tobena-Santamaria, R., Laxalt, A. M., Jones, D. R.,

Divecha, N., Gadella, T. W., Jr., and Munnik, T. (2006) Visualization of PtdIns3P dynamics in living plant cells. *Plant J* **47**, 687–700.

27. Stahelin, R. V., Burian, A., Bruzik, K. S., Murray, D., and Cho, W. (2003) Membrane binding mechanisms of the PX domains of NADPH oxidase p40phox and p47phox. *J Biol Chem* **278**, 14469–14479.

28. Ellson, C. D., Gobert-Gosse, S., Anderson, K. E., Davidson, K., Erdjument-Bromage, H., Tempst, P., Thuring, J. W., Cooper, M. A., Lim, Z. Y., Holmes, A. B., Gaffney, P. R., Coadwell, J., Chilvers, E. R., Hawkins, P. T., and Stephens, L. R. (2001) PtdIns(3)P regulates the neutrophil oxidase complex by binding to the PX domain of p40(phox). *Nat Cell Biol* **3**, 679–682.

29. Ellson, C. D., Anderson, K. E., Morgan, G., Chilvers, E. R., Lipp, P., Stephens, L. R., and Hawkins, P. T. (2001) Phosphatidylinositol 3-phosphate is generated in phagosomal membranes. *Curr Biol* **11**, 1631–1635.

30. Kanai, F., Liu, H., Field, S. J., Akbary, H., Matsuo, T., Brown, G. E., Cantley, L. C., and Yaffe, M. B. (2001) The PX domains of p47phox and p40phox bind to lipid products of PI(3)K. *Nat Cell Biol* **3**, 675–678.

31. Levine, T. P., and Munro, S. (1998) The pleckstrin homology domain of oxysterol-binding protein recognises a determinant specific to Golgi membranes. *Curr Biol* **8**, 729–739.

32. Levine, T. P., and Munro, S. (2002) Targeting of Golgi-specific pleckstrin homology domains involves both PtdIns 4-kinase-dependent and -independent components. *Curr Biol* **12**, 695–704.

33. Balla, A., Tuymetova, G., Tsiomenko, A., Varnai, P., and Balla, T. (2005) A plasma membrane pool of phosphatidylinositol 4-phosphate is generated by phosphatidylinositol 4-kinase type-III alpha: studies with the PH domains of the oxysterol binding protein and FAPP1. *Mol Biol Cell* **16**, 1282–1295.

34. Godi, A., Di Campli, A., Konstantakopoulos, A., Di Tullio, G., Alessi, D. R., Kular, G. S., Daniele, T., Marra, P., Lucocq, J. M., and De Matteis, M. A. (2004) FAPPs control Golgi-to-cell-surface membrane traffic by binding to ARF and PtdIns(4)P. *Nat Cell Biol* **6**, 393–404.

35. Balla, A., Kim, Y. J., Varnai, P., Szentpetery, Z., Knight, Z., Shokat, K. M., and Balla, T. (2008) Maintenance of hormone-sensitive phosphoinositide pools in the plasma membrane requires phosphatidylinositol 4-kinase IIIα. *Mol Biol Cell* **19**, 711–721.

36. Roy, A., and Levine, T. P. (2004) Multiple pools of phosphatidylinositol 4-phosphate detected using the pleckstrin homology domain of Osh2p. *J Biol Chem* **279**, 44683–44689.

37. Gozani, O., Karuman, P., Jones, D. R., Ivanov, D., Cha, J., Lugovskoy, A. A., Baird, C. L., Zhu, H., Field, S. J., Lessnick, S. L., Villasenor, J., Mehrotra, B., Chen, J., Rao, V. R., Brugge, J. S., Ferguson, C. G., Payrastre, B., Myszka, D. G., Cantley, L. C., Wagner, G., Divecha, N., Prestwich, G. D., and Yuan, J. (2003) The PHD finger of the chromatin-associated protein ING2 functions as a nuclear phosphoinositide receptor. *Cell* **114**, 99–111.

38. Karathanassis, D., Stahelin, R. V., Bravo, J., Perisic, O., Pacold, C. M., Cho, W., and Williams, R. L. (2002) Binding of the PX domain of p47[phox] to phosphatidylinositol 3,4-bisphosphate and phosphatidic acid is masked by an intramolecular interaction. *EMBO J* **21**, 5057–5068.

39. Zhan, Y., Virbasius, J. V., Song, X., Pomerleau, D. P., and Zhou, G. W. (2002) The p40phox and p47phox PX domains of NADPH oxidase target cell membranes via direct and indirect recruitment by phosphoinositides. *J Biol Chem* **277**, 4512–4518.

40. Dowler, S., Currie, R. A., Campbell, D. G., Deak, M., Kular, G., Downes, C. P., and Alessi, D. R. (2000) Identification of pleckstrin-homology-domain-containing proteins with novel phosphoinositide-binding specificities. *Biochem J* **351**, 19–31.

41. Manna, D., Albanese, A., Park, W. S., and Cho, W. (2007) Mechanistic basis of differential cellular responses of phosphatidylinositol 3,4-bisphosphate- and phosphatidylinositol 3,4,5-trisphosphate-binding pleckstrin homology domains. *J Biol Chem* **282**, 32093–32105.

42. Kimber, W. A., Trinkle-Mulcahy, L., Cheung, P. C., Deak, M., Marsden, L. J., Kieloch, A., Watt, S., Javier, R. T., Gray, A., Downes, C. P., Lucocq, J. M., and Alessi, D. R. (2002) Evidence that the tandem-pleckstrin-homology-domain-containing protein TAPP1 interacts with Ptd(3,4)P$_2$ and the multi-PDZ-domain-containing protein MUPP1 in vivo. *Biochem J* **361**, 525–536.

43. Xu, C., Watras, J., and Loew, L. M. (2003) Kinetic analysis of receptor-activated phosphoinositide turnover. *J Cell Biol* **161**, 779–791.

44. Sun, Y., Carroll, S., Kaksonen, M., Toshima, J. Y., and Drubin, D. G. (2007) PtdIns(4,5)P$_2$ turnover is required for multiple stages during

clathrin- and actin-dependent endocytic internalization. *J Cell Biol* **177**, 355–367.

45. Dove, S. K., Piper, R. C., McEwen, R. K., Yu, J. W., King, M. C., Hughes, D. C., Thuring, J., Holmes, A. B., Cooke, F. T., Michell, R. H., Parker, P. J., and Lemmon, M. A. (2004) Svp1p defines a family of phosphatidylinositol 3,5-bisphosphate effectors. *EMBO J* **23**, 1922–1933.

46. Krick, R., Tolstrup, J., Appelles, A., Henke, S., and Thumm, M. (2006) The relevance of the phosphatidylinositolphosphate-binding motif FRRGT of Atg18 and Atg21 for the Cvt pathway and autophagy. *FEBS Lett* **580**, 4632–4638.

47. Yan, J., Wen, W., Xu, W., Long, J. F., Adams, M. E., Froehner, S. C., and Zhang, M. (2005) Structure of the split PH domain and distinct lipid-binding properties of the PH-PDZ supramodule of alpha-syntrophin. *EMBO J* **24**, 3985–3995.

48. Venkateswarlu, K., Oatey, P. B., Tavare, J. M., and Cullen, P. J. (1998) Insulin-dependent translocation of ARNO to the plasma membrane of adipocytes requires phosphatidylinositol 3-kinase. *Curr Biol* **8**, 463–466.

49. Varnai, P., Rother, K. I., and Balla, T. (1999) Phosphatidylinositol 3-kinase-dependent membrane association of the Bruton's tyrosine kinase pleckstrin homology domain visualized in single living cells. *J Biol Chem* **274**, 10983–10989.

50. Campbell, R. E., Tour, O., Palmer, A. E., Steinbach, P. A., Baird, G. S., Zacharias, D. A., and Tsien, R. Y. (2002) A monomeric red fluorescent protein. *Proc Natl Acad Sci USA* **99**, 7877–7882.

51. Fivaz, M., and Meyer, T. (2003) Specific localization and timing in neuronal signal transduction mediated by protein-lipid interactions. *Neuron* **40**, 319–330.

INDEX

Christopher J. Barker (ed.), *Inositol Phosphates and Lipids: Methods and Protocols*, Methods in Molecular Biology, vol. 645,
DOI 10.1007/978-1-60327-175-2, © Humana press, a part of Springer Science+Business Media, LLC 2010